THE VESTIGIAL HEART

THE VESTIGIAL HEART

A Novel of the Robot Age

CARME TORRAS

Translated by Josephine Swarbrick

THE MIT PRESS   CAMBRIDGE, MASSACHUSETTS   LONDON, ENGLAND

Originally published in Catalan as *La mutació sentimental* by Pagès Editors, © 2008. Winner of the X Manuel de Pedrolo Award.

For additional reading materials and discussion questions, see https://mitpress.mit.edu/books/vestigial-heart

Epigraph from *The Passions* by Robert C. Solomon, © 1993. Reprinted with the permission of Hackett Publishing Company.

The quotation on page 47 comes from *Im Schatten Albert Einsteins: Das tragische Leben der Mileva Einstein-Marić* (In the shadow of Albert Einstein: The tragic life of Mileva Einstein-Marić) by Desanka Trbuhović-Gjurić (Bern and Stuttgart: Verlag Paul Haupt, 1983).

This book was set in Sabon by Westchester Publishing Services. Printed and bound in the United States of America.

Library of Congress Cataloging-in-Publication Data

Names: Torras, Carme, author. | Swarbrick, Josephine, translator.
Title: The vestigial heart : a novel of the robot age / Carme Torras ;
 translated by Josephine Swarbrick.
Other titles: Mutació sentimental. English
Description: Cambridge, MA : The MIT Press, [2018]
Identifiers: LCCN 2017038907 | ISBN 9780262037778 (pbk. : alk. paper)
Subjects: LCSH: Robotics--Fiction. | Emotions--Fiction. |
 Bioengineering--Fiction. | Memory--Fiction. | Science fiction.
Classification: LCC PC3942.43.O763 M8813 2018 | DDC 849/.936--dc23
 LC record available at https://lccn.loc.gov/2017038907

10 9 8 7 6 5 4 3 2 1

"It is the relationships that we have constructed which in turn shape us."
—Robert C. Solomon, *The Passions*, 1977

CONTENTS

I

ROBOTS, MASSAGES, AND AN ADOPTION

1

ALPHA+

7:10 a.m. – I observe Dr. Craft's restless sleep. He is snoring. I move closer to the bed and connect the microphone to the medical channel. Are you reading the snores? Recording confirmed. I attach the report: he is lying on his left side, tonight he turned over twenty-nine times and got up twice to urinate, he had twelve episodes of apnea between thirty and fifty seconds each, now he maintains thirteen inhalations per minute and a regular pulse of sixty-two. I request approval to raise his dose of Rhinofor and end the communication.

7:15 a.m. – Time to initiate the tactile alarm. I combine moving the duvet lightly with rubbing his cheeks and forehead. In the new sequence I correct what he did not like yesterday: I don't even think about touching his nose, and insist more on the eyebrows, moving slowly down across the temples. He opens one eye, groans a little and rolls over to face the other way. I go round the bed and start the tactile sequence again while playing the melody, tailor-made for the Doctor, that I have downloaded from the central repository to stimulate his emotional reserves for the day.

"Get off me, you confounded beast," he bellows, giving me a shove that my joints have no problem absorbing. A good sign, he liked it. I must reinforce this sequence.

7:20 a.m. – I repress the primary programming to offer him my arm to help him up. Every morning millions of ROBs all over the world make this gesture toward their PROPs, but I have to inhibit it. The Doctor is a rebel and I have to adapt myself to him. I have been built with a boosted learning capacity precisely because he is difficult. And he is the boss. Neither can I tell him that if he gets out of bed on the left-hand side, after only two paces he will be in the bathroom. I have to let him go all the way round. He knows this is the longest way, but "Why make things easy if you can do them the hard way?" he let fly at me one day. It is not logical. He says he likes the "re-creation" he gets out of things, and he pronounces it like that, really separating the "re." I analyze everything he says, to adapt myself to him as much as I can, but the return is low. About the only thing I have learned so far is to inhibit my primary reactions.

7:25 a.m. – I never go into the bathroom when he is in there. Another "no" I had to learn. But I connect to the toilet to receive the analysis. The first urination of the day is the most important. pH = 6, negative proteinuria, slight glycosuria. Attributable to last night's excessive inebriation. Fecal analysis: microbial flora parasite-free, leucocytes within limits. I transmit: normal excretions, low-sugar diet recommended for today.

DR. CRAFT

The man sitting on the toilet looks at himself in the mirror and barks. My face is starting to look like a dog's, he thinks, and barks again. Does that happen to all old people? Tonight, taking advantage of the face-to-face soirée, he will pay close attention to this. Good night, Mr. Bulldog; come on in friend Fox-Terrier, your husky is looking splendid this evening. Nice to meet you, yes, I'm Dr. Pit-Bull. Urrrggg. If he doesn't tense up his face, everything hangs down: the bags under his eyes, his formerly fleshy cheeks, his jowls; by instinct he lowers his gaze, to look at his belly, the muscle-less skin on his thighs, his cock. Look at it, so inoffensive. And it got to be so despotic, the bastard. First always chasing women, and then, that bloody prostate, it didn't let him live in peace. Getting old has its advantages: now it is he, and only he, who makes decisions, every hour, every second of the day.

Alleviated, he gets off the toilet and stands up facing the mirror. He frowns and a few black hairs, long and untameable, shoot up, giving him a diabolic look that reconciles him with his physical appearance. Urrrggg. Quite an idea, that one about the dogs. Maybe we could even get a product out of it: "Youngster, want to see yourself thirty years from now? Turn on your ROBcam and stand in front of it." Photographs of the future. One would just need to select a breed of dog based on the most characteristic traits of the face and morph the youngster's face with that of the dog. This evening he'll try it with Hug 4'Tune and Fi. Without any warning, he'll project the images directly onto their mirrors. What a fright they'll get when they suddenly see themselves looking so old.

He lets himself fall backward into the immense bathtub and the sponge net catches him and rocks him like a baby. It doesn't even occur to him that it could fail. Up and down. Down and up. The waves break against the ceramic walls creating lines of bubbles that burst against his body. The tickling makes him shudder but, once he gets used to it, his other senses are opened to the revitalizing fragrance of a sunny morning and the cheering melody that is still playing. Ebullient. He feels ebullient. Alpha+ has arranged everything to perfection. He would trust Alpha+ over and above his mother, if she were still alive. Or, it goes without saying, his daughter or his wife. The robot was already a good servant, but since it had the neuroaccelerator installed it is learning at a vertiginous speed, and in a few days has adapted to him like a tight glove.

Like this water, which also molds itself around every part of my body, he thinks, as he moves his legs so that the warmth fills his most intimate nooks and crannies. A good choice of stimuli, that's the secret to well-being. Let's forget about self-help implants and other neuropsychological devices, we can't change man or turn his brain upside down, we can't modify even the smallest reaction. Let's accept that. The only way forward is to control his surroundings, control what he feels through the stimuli he receives. A key idea, but when he presented it as the leitmotif of the new line of robots, no one gave it a bit's worth of notice. Too simple, they said. How short-sighted! One must understand man, each man, in order to be able to activate the right resources at the right time. This was the difficult part: they couldn't tailor-make a ROB for everyone; they had to come up with a generic ROB that was highly adaptable and, most important of all, one that could achieve a very fast adaptation. If it took one week for a ROB to work out how to wake up its PROP or how much sugar to put in his coffee, the whole idea would go down the drain. But he was sure: at CraftER they had the expertise to do it. Their competition didn't; that's why he rubbed their noses in the idea. They couldn't plagiarize it. The only thing that made him occasionally stumble was the speed, for a very long time it had escaped them ... and now Alpha+ is the proof that he was right, the culmination of his idea.

The sponge net draws back to the bottom of the bathtub, just as his body calls for him to perform some swimming strokes. He takes a deep breath and submerges himself; he exhales, outward, inhales, inward. The smell of eucalyptus fills his lungs and gives him the impression that he is moving forward more than ever with each push. He empties his mind, abandoning himself to the pure sensation of his own strength, gliding through the water.

When he gets tired, he stretches out face up and lets himself float, and the net picks him up again, cradling him gently. If we could do the same thing with the mind ... At its heart physiology is easy: trying out stimuli and measuring reactions, that's all. We could even control feelings in this way: "Hey, Hug, go and see Fi, her emotional state has just become compatible with yours." If it weren't for the fact that we can't play with person-stimulus so frivolously, because ... who would be given priority? With compatible states, mutually stimulating, there wouldn't be any problems. Nor with incompatible ones. But what would we do when one person wants it and the other doesn't; when for one it's harmful and for the other ... Like the blood groups: O- people, universal donors, ultimate altruists, can only receive blood from those of the same blood group. At least at the level of blood everything is organized into groups, something that's unimaginable for feelings, an inextricable network of attractions and rejections. Even so, it would be good to have a panel with an LED for every person you know. If the light is on, their emotional state is compatible with yours; if not, it would be better not to get too close. Each person could make their own choice, on site ... or it could be done centrally, to guarantee the maximum global satisfaction. What an invention, the electronic matchmaker.

He had been so absorbed in his brilliant reflection, and all of a sudden he's wondering whether it might be nonsense. His thoughts had never embarrassed him before, but now it happens often. As he gets older, he's becoming like other people, clichéd and banal, that's how his mind will end up. He's still capable of following anyone's logical reasoning, but when he lets his imagination fly he gets caught up in developing theories of the most ordinary nature. Over the years, his brain has lost its originality. And that used to be his strong point, what catapulted him above the run-of-the-mill engineers.

So many prostheses for everything nowadays, and there's not even one for his handicap. Damn the LED contact panels! What he wants is a creativity prosthesis. Or an assistant, it doesn't matter; something that would stimulate him to think differently, that would warn him when he started down well-worn paths and would show him the promising forks in the road, those susceptible to innovation. Now the net has placed him on the massager at the side of the bathtub and a series of cushioned rolls and strategically placed heat sources are drying and massaging him from head to toe. A brain massage, that's what he needs.

2

LU

Lu's had a hard time finding clothes like the ones they wore at the start of the twenty-first century. Luckily for her the fashion was revived only fifty years ago, because on this point the psychologist was categorical: it was absolutely essential that she dressed like that so as not to shock the little girl. With some difficulty she pulls on a pair of blue trousers so hard and rough that she shudders, and an off-white t-shirt that makes her wince with disgust. She couldn't find leather shoes, and hoped the girl wouldn't notice. She spies herself in the mirror out of the corner of her eye, but tries not to look too closely, and the image of her grandmother smashes into her retina. Today she does bear a resemblance to those old photos, void of depth. She'd hated them so much as a child, but the old woman insisted on looking at them on her relic of a computer. Look sweetie, we're so alike, can't you see? But Lu was horrified by the black-rimmed eyes, the sagging cheeks, and, above all, those tight trousers that opened up like a glass onto a bloated belly without any trace of waist.

Not even remotely, Grandma, we don't even look alike when we're dressed in the same old rags. She carries on looking at herself as she turns on her toes, and thinks that the passive gymnastics and massages have done a megabyte's worth of good! Thanks to ROBul, her prominent hips taper into a model's waist. But how will she get by now? How will she survive without her robot? She's proud of herself for being able to accept such a great sacrifice. They've assured her that it will only be a couple of weeks, until Celia feels secure in her new lifestyle. Celia. She stares at the bedroom door and tries to imagine her there, first wary, not daring to look at her, and then running toward her. Throwing out her arms and letting herself be drawn in close. In a new gesture that she enjoys, Lu hugs herself, with her eyes tightly closed, and she can almost feel the girl's breath against her stomach.

She slowly drops into the chair beside the dressing table and imagines how lovely it will be to comb that beautiful hair, so straight, and longer than she's ever seen. The doctors said that it would take a good three years for it to fall out, but no matter how much she insisted on the subject she couldn't get a clear answer on why the hair loss couldn't be avoided. They only came up with excuses like oh well, it's all down to environmental damage causing genetic mutation, oh, in the last few decades the evergreen trees have become extinct as well. They go on and on with their big words just so they don't have to admit that they're completely ignorant, those know-it-alls, always so full of themselves. And then that old joke that bald people are delighted every time autumn comes around and all our hair falls out...girls' too, one of the doctors had added, contemplating the eight months' worth of hair on her head. Morons. She'll take it upon herself to look for a decent specialist and everyone will be jealous of her Celia. Starting with that idiot Fi, for all that she labeled adoption frivolous and inhumane, for all those ironic comments about how now that the third world has been eliminated we have to import kids from the previous century, even she will want to be in Lu's position.

She pulls her white-blonde hair back off her face, and goes about putting her makeup on. So many years have passed since she last did it herself that the products have evolved a gigabyte. Now they're intelligent, and can both improve your mood or hide it, while simulating the polar opposite of it. It's important not to make any mistakes, because after a certain amount of time the simulation can impose itself and end up completely changing the situation. Luckily, ROBul has carefully prepared it all and there are labels and instructions for everything. She selects several options from the dressing table panel. *Partenaire:* daughter; empathy: high; situation: welcome; mood: cheerful; emotion: tenderness. There's no need to fill in the date, time or climate, because the program detects them automatically. So she presses the button indicated by the arrow, and the chosen concoctions for face, eyes and eyelashes gently place themselves on the table. She realizes straight away that she no longer has the knack for putting her own makeup on and the whole process drives her mad, but it's just the price she has to pay, and it'll only be for two weeks. Once she finishes, a youthful and maternal Lu smiles back at her from the mirror. "Sertuum—we work miracles," the fleeting ad reminds her, superimposed under her face like a necklace.

On the way out Lu goes to cover herself with her anti-UVA hood as usual, but the contrast with her current attire makes her stop for a moment, as if the psychologist had caught her in the act. An old-fashioned coat would be enough to combat this Nordic cold, it must be more or less the same as before, but her skin and hair would be damaged by all the radiation. She can't expose it. Anyway, there's no way the little girl will see the cover up, it will be left in the decontamination chamber at the entrance. That was the only thing she liked about the clinic, that glassy cabin, capable of delicately removing her protective clothing, without any effort on her part, and promptly returning it to her on the way out. She had seen similar cabins in other medical centers and health clubs, but none were as elegant as this one. That was where the up-to-date technology ended, at least at the visible level, because the rest of the clinic was to all appearances megabytes old, so as not to scare the kids, as they hurried to tell her. It takes some balls to justify deficient service with such a puerile excuse. The same excuse that, according to them, meant they had to have the exclusive clinic in a deserted area on the outskirts. As if the children would be traumatized if they saw some iron and concrete through the window, and, sure, a few aero'cars, but they'll see plenty once they leave anyway.

At this rate it'll be her that won't make it out the door. She's never had to worry about preparing the house for her return and, because of this lack of expertise, she is stuck in front of the control panel. She misses ROBul more than she'd thought. Even though she's followed his instructions bit by bit, she gets the impression that something isn't quite right.

But there's no time to hang around, the last thing she needs is for the little girl to wake up without her there. She would be refused the adoption, they were very clear about that: to establish a good filial connection the first thing the girl sees must be her new mother's face. Hopefully it will work. Celia, Celia, Celia, the name beats strongly through her, in her head, neck, everywhere. She doesn't like the name, but for now she's not allowed to change it.

3

SILVANA

If they hadn't known her, the people who crossed her path in the corridors of the ComU would have seen a mature woman, getting on for fifty, blonde, with watery eyes and that falsely careless appearance of someone who knows they possess an innate beauty. Everyone does know her, however, and above all what they see in her is the wife of Baltasar, leader of the "Stop the Boomerang" movement, and the reason they're all there. People know she's a woman of character, an experienced fighter, and that no one will ever get a taste of her ... despite her having a smile for everyone, touching a nerve with her contagious optimism, and awakening in more than one person the thrill that anything is possible.

On the way to work, aside from greeting people here and there, she only stops to collect her order from the bibliographic agency. While it hands over two of the volumes she's ordered, the dispensing machine advises her with its metallic voice that it hasn't been able to get ahold of *The World of Yesterday* by Zweig yet, because there aren't any copies in the Public Collection and it's had to ask for the digitalization rights. As long as there are no problems from the owner she'll be able to collect it in a couple of days.

This small inconvenience revives her desire to dive into those old books and forget about the time and day. Or, even better, get the day to forget about her. It's not exactly that she dislikes the emotional stimulation sessions, on the contrary, within an hour she'll be physically and spiritually immersed in it and she won't be able to imagine having done anything better in her whole life. But in this particular moment she gets the impression that Gandhi's biography and, especially, that little book with the suggestive title, *A Man,* by some author called Fallaci, will provide the key to unlocking the feeling that she's spent months searching for, and that, up to now, has proved so elusive. Yesterday, she thought perhaps that feeling was not captured by the concepts and categories she was working with, and that the descriptions of twentieth-century people would help her discover the elements that were lacking.

When she goes into the health center she heads straight to ask Sebastian, the incident manager for her unit, if she can skip the clinical case session this afternoon. Sebastian, the only founding member who has known her since before she got together with Baltasar, puts his arm around her shoulders and invites her to sit down for a while. He wants to make sure that everything's going well and, briefly, to find out what's on her mind.

"Don't worry, there haven't been any emergencies in your area." He sits down at touching distance: his legs are never-ending and his torso is slightly inclined to act as a counterweight. "But we like you to be there, and lately you've not shown up so often."

"I've shown up plenty of times over the years." Her tone is firm without being scornful. There's no way she'll let her afternoon of reading be snatched away now that she knows it's within reach.

"That's a pity. You were the one who pioneered this, the heart and soul of the discussions." Their arms, very close, lightly brush against each other. "You said that dissecting a case, comparing the latest scientific findings with clinical experience, confronting different opinions ... basically, poking holes in one another's arguments"—he pretends he's going to prod her, and instead strokes her breast—"for young people it was the best kind of education, remember?"

"Yes, I still think so. It's just that those young people have grown up, and now it's up to the next generation to take up the gauntlet. I have other challenges now, other projects ..."

"Damn! What a way to talk. Next you'll be telling us you're joining the corporations on the other side."

"I need some time, Seb." She lowers her tone an octave and searches for his pupils with her own eyes. "I've been chasing after the same idea for months and, finally, it seems I'm about to get to the bottom of it."

"You and your dreams, Silvana. In thirty years you've not changed a bit." Now it's him who takes her by the arms and locks his pupils onto hers. "When I look at you I still see the eighteen-year-old girl ready to take the world by storm."

Suddenly immersed in an enchanting musical score, a duet engraved on their bodies on a level beyond thought, they stand up and embrace, their lips soaring toward each other until they meet. For a few seconds they spin, little by little, savoring the moment, before they sit back down and resume the conversation as if nothing had happened.

"Of course I'll take the world by storm! Maybe not today's, but the world of a couple of centuries ago." She crosses her legs seductively.

And he takes the bait:

"Go on then, tell me, which new feeling has you in its thrall now?" His hand is on her leg and her reaction is condescending. "'New' is a turn of phrase. Come on, I haven't forgotten that you're only interested in anachronistic emotions, and better still extinct ones."

"I'd love to explain it to you another day when I have more time. Today I'm wanted elsewhere." His hand is on top of hers, stroking it. "But I will tell you that it was an affection that required distance; weird, isn't it? Exactly the opposite of what we preach here. Neither beauty nor physical capabilities, neither touch nor any other sensation have anything to do with it. And don't get the wrong idea, it's got nothing to do with an electronic connection either." She's so serious that he instinctively withdraws his arm and attempts to meet her on an intellectual level.

"But was it attractive or repulsive? Dominant or … ? Where does it fall on the psychogenetic scale?"

"I've not managed to situate it yet, I'm still trying to work out its exact composition. I'm sure the frontal lobe, the amygdala and part of the limbic system are involved, because there are elements of ambition, respect and a capacity for suffering. It would be categorized as an attraction if it weren't that, oddly enough, it doesn't permit bodies to get too close to each other. Maybe it's because it tended to flow from a younger person to an older one … with some exceptions, of course, which always complicates things. So you see, Seb, it doesn't fit in our frame of reference." She places both feet on the floor, ready to stand up.

"And is there any trace of it left in the current population?" Absorbed in the topic through those magnetic eyes, he resists abandoning the subject.

"Not one. That's why I'm not sure I'll be able to get my head around it without references, context, the right biochemistry, perhaps even without the appropriate organs."

"Has it really been extinct so long that the substrate has atrophied? Is it not susceptible to your stimulation programs?" The questions flow out of him.

"That would be nice, but I'm afraid it won't work. It's got nothing to do with jealousy or honor, which are much closer to us on the evolutionary ladder. Nor with patience …" She stands, mechanically kisses him on the forehead and leaves the room without looking back at him. "Mine has been stimulated too much."

She knows a decent group of candidates is waiting for her in ESZ, desperate, as always, to offer up their skin in order for her to open up new horizons for them with her hands. She used to be so proud, but now she finds herself asking more and more often what right she has to change their lives. They come because they want to, of course, almost certainly unsatisfied with the day-to-day monotony of their work and electronic relationships. They've been told or they imagine that experiencing new feelings will be fun and, no matter how much they're warned, they won't believe that they'll be taking a leap into the unknown. She threw herself into it because of her beliefs, seeing it as a necessary first step on the way to recuperating lost emotions, but most volunteers aren't looking for that. It's hard enough to get them to understand that here "e-motion" doesn't refer to the movement of electrons.

She speeds up to escape the thought that's hot on her heels, the accusation thrown at her before she left the last session by that English girl who was in a dreadful state: "in the old days people put their hands on each other to heal, but you only cause pain." A few months ago she would have replied "no pain, no gain," and left it at that. But now she sees it differently. Not everyone longs for knowledge and it's not fair to make them suffer with little possibility of a reward. Even if they are volunteers. They invest a level of trust in her that is much more than she can accept. Before she used to feel certain of what she did, confident she was familiar with the propagation pathways, the neural mechanisms ... but now she knows that the only thing she actually dominates is the detonator. Once the process has been set off, she becomes an attentive and conscientious guide, that much is true, but without having a lot of control over the deeper dynamics, and less over any bifurcations.

She passes, as usual, through the S in the middle of the enormous holographic ESZ sign that marks the border of the Emotional Stimulation Zone, which takes her straight to where she needs to be. Benjamin has already sorted the participants into pairs, one lying face up and the other kneeling at their feet, fanned out before Silvana. The floor is made of a soft, smooth material that hardly adapts at all to each person's anatomy, making them squirm uncomfortably, searching for a better position. Not one person returns her gaze when she lets it linger on them, trying to work out who will be able to offer her a good response, and who will end up reproaching her for their suffering. No one. Each new batch has less character: they really are lost, it's too much even to ask them where their feet are. She takes the foot of one of the participants who's closest to her and, dexterously pressing on it, she traverses the kid's big toe and then moves down toward the heel in a zig-zag motion. The boy writhes like an eel, Silvana literally has his body's remote control in her hands. When she asks him what happened, he looks at her, disoriented, and tells her nothing happened, he's not aware of having moved a muscle.

For her, on the other hand, the relationship between the pressure points and the resources unleashed was so clear that, in one sweep, she has identified a whole, miniature blueprint of the body, with each major organ marked on it. With a tactile pointer, she indicates the references on his foot and then guides his partner's hand over the same route she used before. From there an adventure game begins in which they must collaborate in a profound exploration of the path, mutually orienting each other, until they've used up all possible variants of this basic exercise.

She takes advantage of the time spent showing each couple the same exercise to warn them:

"Tactile stimulation is only the first step on a long journey ... which can be very painful for some," she adds, remembering the English girl.

The statistics show that half of those here will leave the program after three sessions and only a couple will finish the course.

Poor kids, she feels sorry for them while she's preaching to them, every day they seem more defenseless. To think that amoebas are born adults and deer learn to run and feed themselves within a couple of days, whereas these idiots…they haven't even realized they have a body yet! The Peter Pan generation, as Balt calls them, are already here: the ones that were born among the ROBs and, in actual fact, have been brought up by them. The sensitivity awareness campaigns, massages and emotional stimulation courses won't be of much use to them. We're too late. More drastic action is required to stop the boomerang effect. If not, soon the invalidity phase will last a whole lifetime and people will get old and die without ever growing up. It is true that the longer it takes an animal to grow up, the more complex it is, but there's a limit to everything, and, in the end, the most developed species on the planet will die of over-evolution. Silvana's own grandiloquent thought shocks her, and, focusing her gaze on a belt full of sensors pressing down on the prominent belly of the boy in front of her, she tells herself that they will die, more prosaically, of a glut of technology.

4

CELIA

What's going on? Someone's come to wake her up… or is it just a noise? She's extremely tired and her eyes are so heavy she can't open them. And she can't move either. Mommy! She shouts internally, but her voice doesn't come out and her mouth isn't doing what it's told. Maybe she's dreaming. What happened last night? All she remembers is that they gave her another one of those injections… that's why she's so exhausted. She wants to sleep a little more. Her mom will wake her up when it's time. She's uncomfortable but she can't roll over. At least the pressure in her chest has been relieved and she doesn't have a headache any more… only this fog. She can breathe fine: up, down, up… oh, there's a strange smell here, like cough syrup, it reminds her of Nancy. She was so excited when her dad brought her back from Andorra! Where would she be now? It's been a long time since she's played with dolls, but she still loves her, a lot. She'll ask him to look for her and bring her here again: she'll put her on the bed by her side for company. Ah, her arm has twitched. She tries moving it again, and yes, it twitches! So she's not dreaming. She can hardly open her eyes, with such a bright light, it's blinding her.

“Psst! She’s awake.” The nurse pushes Lu closer to the bed and, with the dexterity of someone who does it often, places Lu’s hand on top of the girl’s, immediately retreating into the background and adding loudly with an overly sweet tone: “Hello Celia, darling, how are you feeling?”

Whose is that hoarse voice? It could be her grandma … but … did she just call her “darling”? No, she calls her sweetie or lamb or … how’s my little girl today? Anyway, she said goodbye yesterday with lots of kisses because she was going back to Gurb: she had to prepare the house for … she hadn’t really properly understood for whom. Whomever it was, she wasn’t very happy about it, because you could see she was upset and, when she said goodbye from the door, her eyes were brimming with tears. Poor Grandma, still having to do things she doesn’t want to at her age. Again the voice and that cold hand. It must be a new nurse … maybe if she pretends to be asleep she’ll escape the injection. Or are they bringing her breakfast already? She’s not hungry at all. Though she’s not nauseous like the other day, her stomach feels strange: it feels like her heart has sunk down to her bowels to stir things up. Maybe she’s getting her period. Her mother says the pain feels very different, unmistakable, but for Celia, the two times she’s had it, she’s not known how to tell it apart from a stomachache.

“Say something to her, go on,” the nurse whispers into Lu’s ear. “I’m sure she can hear us; the signals”—she points toward the monitors—“show she’s conscious.”

“Celia, can you hear me? I’m here, by your side.” She stretches her neck out so that her face is right in front of the girl’s. Several times the psychologist had said it was absolutely fundamental that when she opened her eyes she saw her. It’s written in our genes, he insisted, the newborn becomes attached forever to the first person it sees. Celia’s not an infant, but it’s like she’s being reborn.

Ah, now there are two of them. The assistants who come to change her sheets, probably. But what a strange time to do it. Maybe it’s already midmorning? She definitely hasn’t had breakfast … and where’s her mom? She decides to pretend to be asleep until her mother wakes her up. But that annoying woman won’t stop rubbing her hand. She would like to know what they’re saying to each other in such quiet voices, it seems they don’t want her to hear it. Maybe she’s got worse and they’ve taken her to another room. That gives her the shivers. But her parents would be here, they wouldn’t have left her on her own.

"Celia, don't be afraid, I'm here to help you." She says it more to please the nurse than the girl, but the effect is conclusive: eyes as big as saucers open like black holes in a face that suddenly has panic written all over it.

Who is this scarecrow with the distorted face and straw-like hair? Why is she so close? Her hand is sticky and it's making her shudder. Go away! Go away! She doesn't want to see her. Where's her mom? And that wall…Where have they taken her? What a strange brightness. She's very cold and everything's spinning. She feels like she's going to be sick. Mommy, come here!

"Calm down, Celia, don't wear yourself out." The low, silky voice of the doctor acts like a balm on the girl's face. He comes out from behind the machines so the girl can see his face and his white coat, comes closer and, stroking her forehead, adds, "That's better, relax and listen carefully to what I have to say: the treatment was successful, you're totally cured. The suffering, the injections, having to stay in bed, that's all over…you don't notice any pain anywhere, right?"

"Yes, doctor, my stomach hurts." They're her first words and she says them with the seriousness of the most adult patient who completely trusts in the infallibility of medicine and is ready to give all the necessary information.

"Of course, during the treatment your bowels were stopped and it's taking them a while to get going again. But as soon as you take the serum and start eating again the pain will go away, you'll see."

"And where are my parents?" Now she moves her head from side to side, looking for them.

"It would be a good idea not to move around too much, love." The nurse strokes her hair. "You're still very weak."

"They've had to move you very far away, because cases like yours can only be treated in this clinic." Again the nice voice. "Your parents authorized it, of course, and, so that you felt at home, they prepared this box for you. It's full of photos and toys." Seeing that the girl wanted to open it already, he adds, "During your convalescence, Lu will stay with you to help you with everything you need." He pulls a chair up to the bed for Lu to sit in. "Now, if you like, you can show her what's inside. We have to continue with our visits; if you need anything, press the red button."

Once they've left the room, a sense of vagueness, of perplexity, hangs in the air. Two silent spectators had watched on as the white-coated protagonists abandoned them without warning, leaving them face-to-face, at each other's mercy. All at once Lu has lost her place in the script that the psychologist had so precisely drawn up for her throughout all their interviews. She's even wondering if this adoption was such a good idea in the first place. It's only just begun and already the weight is too much to bear. And Celia keeps on looking at her with her eyes too open, intimidating her.

This woman must be the psychologist, surely. Without the white coat and with that expression that means, "I've got all the time in the world just for you, sweetheart, so we'll talk, right?" What a drag! She doesn't understand why her mother insists on trusting these people, saying that they're as professional as the doctors ... or even more so, she sometimes adds, because having a bright outlook helps you to get better more quickly. Why isn't she saying anything? That *is* weird, psychologists talk and talk and never stop asking questions. Please, no!, now she's going to start taking things out of the box and she'll want me to tell her all about each one. She'd better get comfortable.

"Look what's in here: a pretty doll." Lu grabs hold of it as if it were a life preserver.

Wow! It's Nancy! It looks like they've read her mind, her dad sorted that out so quickly ... My Nancy, how exciting!

She squeezes the doll tight with both arms, still shaking with fatigue, and then places her on the bed next to her and covers her with the sheet. The movement stirs up the dead air between her and Lu and, for a moment, a welcome breeze softens the woman's tense face.

"There's also a box with funny shaped pieces and a board ..." Now she regrets not having dedicated more time to studying the games of that era, as the psychologist had recommended. She has no idea what it's all for. And she is only saved when she spots a cardboard folder, identical to the family album her grandmother guarded like a treasure, at the bottom of the box. "Maybe you'd prefer to look at some photos."

She's not going to try to show me them herself, is she? "Leave them on the nightstand, I'll look at them later." Why is she looking so shocked? What's she said wrong? Oh, that's it, there's no nightstand. "Leave them here then, on the shelf." What is it she doesn't get?

There are too many things Lu doesn't understand. She said to "leave them," ignoring her and transmitting orders to some nonexistent robot instead. As if she weren't there. What's she up to? The girl doesn't know the slightest thing about robots. And "nightstand," what does that mean? A dark coat stand? She can tell they're not going to understand each other ... and these are the first words she's said ... No one told her the girl would speak differently, or could stand up to her and give her orders, or dominate the situation more than herself. None of that had figured in her plans and, suddenly, what had been dawning on her becomes obvious: What if the girl doesn't want her? Will they give her another one?

Just the idea of it scares her, so to get it out of her head she opens the photo album and starts to turn the pages, if nothing else it will win her some time. Most of the photos show open spaces, green and brightly lit, in which Celia appears alongside people with their arms around her, touching her hair or with her sat on their lap. Lu can see herself in the pictures, not outside among the vegetation, that's not possible anymore, but caressing the girl, kissing her, brushing her hair. She's sure she'll enjoy always having Celia by her side, it won't be like that high-tech little dog that her psychomanager prescribed her. Yes, it was sweet, but it made her feel like a useless, boring, old woman. When she looks over at Celia to check, she is met with the girl's hard expression, who is pointing at the platform next to the head of the bed.

"Don't you want us to look at them now? Ok, I'll leave them here, on the hi'plat." She goes back to rummaging around in the box, looking for a familiar game.

If she's a psychologist she must be new ... and pretty shy, nervous even. But that can't be right, psychologists always know what to do. Núria always knew. "Miss, who are you? A nurse?" Now, without meaning to, she's really surprised her. You can't say anything to this woman. Why does she always take so long to answer? She squirms as if she were out of practice, and she struggles to meet your gaze; actually, she still hasn't looked at her. The box has her distracted. Maybe she's looking for the words she has such trouble finding.

"I'm not called 'Miss.' My name is Lu. I'm sorry, I forgot to introduce myself, and no ... I'm not a nurse." She wonders for a moment whether to tell her she doesn't work; they made it very clear that for now she shouldn't talk about herself, they should talk about the girl's life instead. "But, let's talk about you ..."

Now she wants me to think she's nice. She has such little personality, poor thing. Celia doesn't feel like explaining anything to her, she just wants to know where her parents are and when they're coming to pick her up. And, if they can't come here, she'll go to them ... since she's cured, isn't she?

Celia insists and insists some more, and Lu doesn't know how to deal with it. The afternoon seems to be dragging on forever, she's already taken everything out of the box and she can't see the light at the end of the tunnel. If only ROBul were here, he'd know how to manage, if nothing else he would entertain the girl and she wouldn't have to worry about it.

The nurse on the afternoon shift is irritated when she sees Lu scurry off as soon as she goes in. These adoptive mothers are all the same, in such a hurry to have a kid and then, as soon as they get one, they're rushing to get out of there and get some peace. As much as she was in favor of intersecular adoptions when she first went to work at the clinic, now she would be happy to defend the right to unfreeze oneself and die. The parents fail to fulfill their duty and leave the kids to their own devices too often. But she, too, has learned to hide from the responsibility. She observes Celia out of the corner of her eye. The girl's picked up the photo album and she's looking at it intently. Without stopping to think about who she's hiding from, she furtively accesses the room controls and advances the diffusion of the repose emissions a couple of hours. Now she can forget about this room until tomorrow.

The photos have been chosen really carefully, Grandma must have done it. Everyone is in them, even people she doesn't know ... that guy, as tall as a giant, obviously has to be uncle Raimon, but he looks different, older, and on the porch there are some deck chairs she's never seen before. Where would they have come from? The garden is the same as ever though, full of flowers with the steps well swept. "Stop complaining so much and get on with your work," her grandma would say to her, but she won't have to say it anymore, because she'll be happy to help with the sweeping. What's this white thing sticking out of the cover? Ah, an envelope ... "For our daughter ..." She has no idea what's happening to her, it's hard to read and her eyes are closing; suddenly she's really sleepy. She leaves the album on the floor, but the letter ... she wants it close to her, here under the pillo

5

ROBco

Status: deactivated. *Cause*: overload. *Details*: I have been functioning at 100% for three days. My PROP has not stopped giving me maps, searching for information, and saturating me with electromyographic records to process. I have not had any available time to carry out the mandatory updates and I have ignored four forced deactivation warnings. Now domestic tasks have been added to the technical ones: Leo has to catch a flight in four hours' time and I have not yet prepared his luggage. The fifth notification was terminal: in order to avoid irreparable damage, it says. My only option was to disconnect.

Status: testing circuits. *Diagnosis*: pending task list overfilled. *Action*: erase low priority tasks. *Error:* all tasks have maximum priority. The scheduler has collapsed and emitted an alarm signal. I need expansions everywhere: memory, processing capacity, speed…, but my PROP does not heed alarm signals. In critical situations, he interferes directly and changes priorities. This goes against all the factory specifications and security regulations. He must know what he is doing, because this is his field, and they say he is one of the best, but he often skips maintenance, and for several months he has ignored pending updates and expansion notifications.

Status: undergoing repairs. He has already put his hands on me. I note how he rummages in my processing unit, he is always so daring, he has not even taken the precaution of isolating it from the neutronic generator. Hey, now he is touching my memory. I think ... finally! He has decided to put in an extension module. But it is not a standard one. Where could he have acquired it? He will have to install a custom protocol. Today's meeting must be important, then. I had assumed it was just another of the same, one of those they hold regularly at CraftER's Prospective Unit to generate new products. Mixing ideas, they call it. But now I realize I was wrong, it is certainly not an everyday meeting, and that is why he has overloaded me and worked himself like never before. And he has done it alone, without consulting other bioengineers as he always does. What is he up to?

LEO

"Why has it crashed now, the fool? Just now." The young, athletic Asian man comes out of the sensory booth where he's doing some last-minute experiments, goes up to his assistant and stares intently at the screen embedded in its chest. "These primitive models ..." he mutters while he opens the lumbar cover, pokes around here and there, moving cables around and checking the state of the picochips, "even if they don't have an auto-repair function, they could at least be able to diagnose what's wrong with them."

"Beep, beep ... *Diagnosis*: pending task list overflowed."

"Okay, Okay, but that's not much help to me, there are thousands of reasons why that might have happened. You'll have to be more specific." He answers without looking up and with his hands still in among the wires.

"*Alarm*: scheduler collapsed."

"You're giving me the consequence, not the cause." Leo's voice exudes impatience. He doesn't know why he insists on talking to ROBco when it doesn't have the capacity to learn. Besides, he's in a hurry. He'll solve the problem the quick way and then sort it out properly when he gets back from Los Angeles.

While he looks for the memory module he bought for his project but never used and installs it in ROBco, his mind wanders to what will happen this afternoon. He's keen to see Mr. Gatew's face when he shows him the total immersion he's achieved. He knows that he's fond of the physical basketball of the golden age, and that he's a self-declared fan of Michael Jordan, that's why he's chosen the game against the Utah Jazz, where His Airness made five or six stratospheric baskets. He needs to convince him to go into the booth himself: the sensation will be much more powerful. The image of the arthritic manager dribbling, passing one, two, three opponents, faking out the colossal Karl Malone, and then launching himself into the air above the mortal souls and making a galactic slam dunk, makes him smile. But that's nothing innovative, Gatew will proclaim with disdain, it's just classic visual simulation, accompanied by muscular, auditory and tactile sensations; that was invented years ago, there are already products on the market that do this. And then he, with great seriousness, will connect the intracranial stimulation. This is the point, my good man, transforming yourself into Air Jordan himself, experiencing what was going through his cerebral cortex in each and every moment. If only he had access to the authentic encephalographic records of the one and only Jordan! That, however, would be techno-fiction and Dwyane Wade's records do the job very well, don't you think, Mr. Gatew? The man wouldn't even hear him, he'd be sucked in by the events of the game. And Leo would have his moment of glory.

He's ended up daydreaming with the module in his hand. Again. If he doesn't fix ROBco soon he'll miss the plane. While he plugs in the extension and manually checks the connections, his mind goes back to the afternoon's presentation, to the key points he has to mention. He's convinced the product will be a success. It's nothing like the limited virtual reality that's been available until now. He's added an intimate perception to the classic sensorial stimulation, the feelings of the star brought to life using a replica of his cerebral activity. And he hasn't yet done away with the idea of eliminating the booth, along with the cables and haptic devices. He'll reduce everything to wave-based communication, no wires, and the user will be able to enjoy the system in any situation—lying in bed, for example. It won't need light projections or recharging or anything. The stimuli won't be directed to the eye, ear or sense of touch, instead they'll be injected directly into the brain. Electroencephalic waves traveling through the air. A dream. Truly a revolution.

ROBco is operative again and is getting everything ready for the trip. It packs the suitcase for a type C2 transoceanic displacement while connecting to the server of the American consulate and requesting the passkey, a procedure that is completed in bursts because, in the middle of it all, it has to answer a series of calls. Using a faithful representation of Leo's voice and intonation, it confirms his attendance at a chess match next Sunday and also the seat reservation on the flight. Bet, however, is not so easy to please, and, once she finds out that Leo is in the sensory booth and that he's leaving in a couple of hours, she demands to speak to him directly.

The young man makes her wait for a while before connecting up the telephonic sound from inside the booth.

"Sorry, Bet, but I really don't have much time. What is it?" As he talks he carries on manipulating the screen and pressing buttons. "Yeah, in the end the manager agreed to see it… but I can't talk about it right now, company rules, I already told you. No, not at the weekend either. I'm sorry Bet, don't keep going on about it." He's stopped working and is rubbing his forehead like he has a headache. "Of course it's not your fault you're working for the competition! Do we have to talk about this now? I thought you were calling to wish me luck." He looks at his watch impatiently. "Stop feeling sorry for yourself, you're having a good enough time swapping gossip about our supercompanies. You're better informed than the two CEOs put together." He pauses for a moment and announces with conviction: "But I can't talk to you about this product… for now." Little by little his eyes open wider and wider and he tries to interrupt once, twice, three times. "Behind your back? What's up with you? It's my work and that's that, I'm not hiding anything. Don't go overboard, okay?" He stands up and sits back down. "I'm absolutely not compromising our happiness app project! They're completely different ideas. As if I wouldn't be able to come up with more than one original product! And our app is much more ambitious, I'm telling you." He stresses this with both hands. "Go on, wish me luck…" He devotes all his attention to her while scratching the back of his neck. "Whatever, but don't come running to me later saying you want me to be successful and all that." A pause. "No, I'm not angry; I'm just in a hurry, I told you that. Yes, when I get back I'll come over to your place. Do you want anything from Los Angeles? All right, I'll bring a couple of bottles. See you on Friday, then. Bye."

6

Finally, they've left me alone! She knows these women have the best intentions but they're all over her, and she's fed up with holding it all in … tears, worries, the sadness gnawing at her insides … and having to pretend nothing's wrong so they don't keep going on and on: you should be happy you're cured, you'll feel better and better every day, you'll see, you'll be back to normal in no time … she can't take any more. The illness has taken away all her spirit and left her empty, as if her body were just a shell that doesn't obey her orders, she has no energy to do anything. She closes her eyes and imagines herself diving down into the dark, stagnant waters inside the shell. Two rays of light penetrate her eye sockets, helping Celia get her bearings, like car headlights on a foggy night. She has to find the nest of sadness and find the energy to fight it by any means necessary. Good spirits, her mother says, which to her sounds similar to the soul. She doesn't know where it is, but she imagines it nestled in her lung, because of her breathing troubles. Her breathing stirs up a current that shakes the diving suit this way and that, and she lets herself be carried by it, without resisting the swaying motion of the salty, dense water. Until, with her back to the headlights, she's drawn into the cavernous corridor of her left leg and, grabbing hold of the blue and red coral tree that is her backbone, she goes in. Swimming forward she leaves the dome of her knee behind her; at the end of the tunnel, coming out of the curve of her ankle, she finds herself in front of the segmented cavities that are her toes. It's hard to get into the little toe, the ebb of the water pushes her back out, and when she manages it, she imagines the feeling of rubbing up against the walls of the cave and it provokes a tickling sensation in her toe that brings her back to her giant body lying in a hospital bed. For a few moments she'd escaped the heaviness and had felt well, like before, happy and light under the water. She wants to go back and hangs onto the image of the crisscrossed veins and arteries ... she can almost see them in those immense, glossy sheets of paper that hung next to her school desk. What a joy it would be to sit there and listen to the singsong voice of Ms. Dalmau. She didn't know it was a pleasure when she was there, or that she would miss it. She wants to go back there right now. She's cured, isn't she? Her parents had promised. Her parents ... Suddenly, all the water she's been holding in this entire time spills over in a river of tears that soaks the pillow.

She cries and cries for a long time, silently, without sobbing or writhing around. She's not looking for consolation, or even to relieve her sorrow, she just abandons herself to the extreme desolation she's feeling.

When she turns over the pillow looking for a dry patch, she finds the letter she'd left there last night. How could she have forgotten? She must be in worse shape than she'd thought, she's lost control of everything, body and mind. She's going mad. Tears spring up again from somewhere deep inside her and she doesn't know if the letters are blurred from the damp pillow or because her eyes are once again brimming over. She dries them as best she can and reads "For our daughter Celia." Her vision clouds over and there's no hope of being able to read, so she's almost pleased when a determined Lu enters with a bunch of huge violets.

"They're like the ones from your grandma's lawn, the ones in the photo. Do you like them?" Celia moves closer to smell them, but Lu immediately withdraws them. "These aren't for eating, only to make the place look nice."

"Yes, of course, what do you mean? I only wanted to smell them." Now she's really confused.

"Smell them? But only animals do that."

"Maybe here, but where I live we really like the smell of flowers." She says this with pride. "These ones don't have much perfume though. Are they natural?"

"They're organic, if that's what you mean."

Lu doesn't even think about sitting down. She puts the flowers next to the photo album, which, she announces, they'll look at later, before theatrically kissing her on the forehead and telling her that she has to go speak with the doctor, but she'll be back in a second to devote herself entirely to Celia, so she should think about what she wants to do.

She's already left the room when Celia manages to recover her train of thought. She's stunned. Núria never kisses her even though they get on really well and they've known each other for months. She's a psychologist, not her aunt or her grandmother. No matter how much she goes over it, she doesn't understand what's going on. Mostly, she doesn't understand what business this woman has here, in a hospital that's so good they managed to cure her. They should have a better psychologist... you don't have to study much to go around kissing people. But she liked it, anyway, and now she wants her to come back and give her another kiss and perhaps a hug. She needs to be fussed over a bit. She's fed up with being in bed.

When she turns over, she once again finds the letter and now she does feel strong enough to read it. She pauses for a moment stroking the envelope between her fingers, but it's not the soft dampness of the paper that gives her the shivers. There are two sheets filled up with her mother's small, regular handwriting, and a postscript of two lines with her dad's smiling sign-off. She likes the face that he's drawn right next to his name, with its big mouth smiling ear to ear. It looks like him as well, with a bald spot, round glasses and a funny expression. She lifts her gaze to what he's written above it and she realizes it's all in capitals letters. Poor Dad, he has such a complex about people not understanding his handwriting. Doctors, he says, have to write the same thing over and over again so many times that their words melt into a deformed line out of boredom and, later, without meaning to, they can no longer define them properly. "My daughter, live life to the full. If you're happy we will be too. We'll always be there with you. I love you." She doesn't recognize her father in those words, he never speaks like that. Of course, he hasn't exactly written her many letters. When she goes away to summer camp it's always her mother who writes. More than anything it's the solemn tone that unnerves her, as if he really were very far away and it would be a long time before they saw each other.

She hastily turns the pages to start reading from the beginning: "Celia, darling, how are you? I hope they've made you better and you're glad you can live a normal life." She doesn't understand what they mean about a normal life, she's still in the hospital. Her mom is always like that: bold words that leave you wondering where they'll lead. Maybe if she carries on reading it'll make more sense. "I don't know how they'll have explained it to you, but I'm sure that by now you understand the decision we've taken." Of course she understands that they've done all they can to get her cured, it's just they could have warned her first. "We would have liked to wait until you were a bit older or, at least, to be able to explain it to you ourselves, but there wasn't enough time and the doctors have told us that knowing about it beforehand could negatively affect your progress." And separating her from her parents doesn't? She has a lump in her throat. The letters she got at summer camp were cheerful, full of words of encouragement from her mother, writing that when she came back they would have loads of things to tell each other and that there would be a party. She's not sure where the sadness that once again blurs the letters is coming from. "It was very difficult to have to choose between your death and a life for you without us." Her tears have dried suddenly, and now she's reading so avidly she's hardly breathing. "You're an intelligent, brave girl, tenacious and passionate about the things that interest you, and you have many interests. Your father and I have discussed it at length: maybe this great curiosity you feel about everything will be your most valuable tool. You have the opportunity to live ahead of your time, so many people would like to, and almost certainly surrounded by fascinating advanced technology." Where have they sent her? To another planet? "Think of it as an adventure and make the most of it. People are desperate to travel far away. You can do it by traveling through time, which is even more exciting." She doesn't even stop to take a breath, she doesn't want to think, first she has to make it to the end. "Often, when you went to school or on a Scouts trip, I told you that I would love to see what you were doing through the hole in my ring, well, now more than ever. So here you have it." The lump in the envelope, she knew it was there but she hadn't stopped to pay it any attention. "When you wear it, it will protect you like an amulet, and it would really make me happy if, from time to time, you took it off so I could see you. In the moments when you feel sad or helpless, you just have to look through the hole: you'll see things close up and then further and further away, and there at the end you'll find me watching over you and comforting you; because when someone is sad they think it will last forever, and that makes them even more sad, but it's not true; the next day you always see things differently. Most of all, don't get hung up on it. I like that you're an introvert and that you think about things, but it's good for you to go out and make friends. Keep your chin up, girl! Wherever we are and wherever you are, you can be certain we're right by your side. I love you more than anything else in the world, Celia. (I won't say goodbye because I'm staying with you.)"

She absentmindedly tries on the ring and twists it around her finger. She's confused. They seem to be telling her she's traveled to the future, but that sounds like a science fiction film. Her parents wouldn't joke about a thing like that. She looks at the walls of the room as if she was seeing them for the first time: they're covered in screens, controls and other devices, nothing out of the ordinary. Perhaps she's never seen so many and all stuck together, but they had told her this was a very advanced clinic. The bed seems normal, too, and the white sheets are made of some coarse fabric. She sits up with some difficulty and swings her legs around toward the floor, but her head starts spinning and she lies straight back down again.

When it seems like she's not dizzy anymore, she tries to reread the letter but, toward the middle, a question hammers so insistently at her brain that she has to stop. Her parents ... are they dead? Just thinking about it makes her heart beat with such force that she feels like she's going to die then and there.

The machines register the alteration and immediately the nurse enters.

"What's going on?" She's looking at the monitors rather than at the girl. "Where's that lady?"

Celia doesn't know who that lady is, nor does she have the energy to think it over, it was hard enough managing to hide the letter under the sheets. Making a great effort she asks what day it is, and, shyly, her voice hardly making a sound, she adds, "and what year?" The nurse's angry look leaves her frozen, before she realizes it should be directed at Lu, "the lady," as the nurse keeps calling her. It's her who's supposed to answer these questions, and it must be against the rules for her to leave Celia alone. She'll have to tell the doctor, the nurse says, as she brusquely storms back out of the room.

That's what she wants most, for them to leave her alone, but at the same time she longs for some familiar hands and a friendly voice to reassure her that it's all just a dream, that soon she'll wake up with her parents at her side. The evasive response has left her with an entirely uncertain future. Or maybe she's dead and this is the afterward that she had worried about so much on those days she got worse and they transferred her to the ICU. But the ring is here. She wraps her fingers around it to form a peephole and focuses far, very far away out the window. A section of electric blue sky hits her retina and she feels that, from out there, her mother is encouraging her to keep going. An adventure, think of it as an adventure, she wrote.

And she has no tears left.

7

10:17 a.m. – Insistent request for an encrypted connection. Checking the key: it corresponds to Mr. Gatew's ROB. We've been waiting for the list of preselected candidates for the E-Creative project for three days. I assign maximum priority. I open the connection straightaway and download the confidential reports.

10:18 a.m. – I go to communicate it to the Doctor, who is seated at the dueling table with his back to me, when an internal alarm stops me in my tracks. I anticipate what he will shout at me: "You moronic heap of scrap, how dare you interrupt my mind in mid-inspiration? Before you start up your synthetic rigmarole, look at this, and you'll be struck dumb by the blow I'm about to dole out to Hug 4'Tune. Never interrupt me again, do you hear me? Never again, you brainless heap of fucking scrap." I have a very accurate model of my PROP, and I know that if I talk to him from behind his back he'll react like that.

10:19 a.m. – The learning module orders me to silently advance toward the armchair, move around it and stand right in front of the Doctor. Done. Now I must carefully observe him while he draws scrawl after scrawl on the horizontal screen set into the top of the table, hoping that he pauses so I can pass on the message. He's so deep in concentration that he hasn't even seen me. Today it's riddles. And it's going well, the scorecard shows that he's two points up. Let's see what he's reading that's putting him on edge.

"In an enclosed monastery, where communication is prohibited and the monks only see each other at lunch time, a message is received informing them of an incurable and highly contagious disease, which reveals itself in the form of a red mark on the forehead. In order to save the community, the diseased monks commit suicide as soon as they know that they are ill. After a few days and a few suicides, the disease is completely eradicated without anyone dying unnecessarily. The question is: how do the infected know that they have the disease? There are no mirrors in the monastery, or lakes, or any reflective surface where the monks can see themselves. Three points for a correct response within half an hour."

Clever little brat. The Doctor has to recognize that 4'Tune has a talent for riddles, no one makes them up like he does. One might say he had a privileged source of information, a programming manual from a century ago, for example, which would explain the profusion of anachronistic elements and the peculiarly algorithmic taste his solutions tend to display. But he's never admitted it. And since copying isn't against the rules, why should he bother lying? What should he care, he's a good adversary, the best in this discipline, which forces him to sharpen his mind. That's what matters. And if, like today, the battle is a high-water mark and he comes out on top, then it's ecstasy. He'll have enough reserves of endorphins to last four days.

He repositions himself in the chair and when he looks up he finds Alpha+ standing stock-still. Hallelujah! He's learned to wait and not interrupt. More than his grandson will ever learn.

"Go on then Alph, what is it?"

10:22 a.m. – Friendly words, no swearing: I reinforce the last intervention to the maximum level on my learning module. I emit: "Mr. Gatew communicates that he has preselected four women and two men, of very different profiles, so that you can choose with total subjectivity, just as you asked. He has sent a confidential report for each candidate and the access path to their record on the public register. He is waiting for your reply in order to submit those chosen to Dr. Cal'Vin's neuropsychological filter."

This news forces the Doctor to postpone the battle. That really was a great advance, the timeout button. Before he would simply have had to abandon the game, just when he was winning. But now this invention can deactivate the encryption database and the riddle will be erased from his memory until he decides otherwise and, most importantly, a certification will be sent to his adversary. Hug 4'Tune will rant and rave as ever as he doesn't have the button, it's a question of privilege, but anyway first things must come first and the creativity prosthesis takes precedence over what is purely a mental game. With the prosthesis, the neurons are in play. 4'Tune can go fuck himself!

Alpha+ supplies him with the documentation as and when he asks for it. Two of the female candidates have renounced their right to privacy, and in the public register there's a link where you can watch them on a live feed. Although he immediately discards their applications since his project is top secret, he connects to the feed. One is captured in a health club, having sex in order to maintain her musculature. If the brain was as fit as the stomach, he would have to hire her, he'll go back to the feed later. The second shows a strange profile, she must have the camera stuck to her foot. No, that's it, she's wearing more than one camera and the image is a mosaic of her body. It's interesting to see her from the sole, in profile and from above, all at once. It looks like an animated Picasso painting... but surely she doesn't know that, she might not even know who Picasso is. He has a look at her CV but there are no clues there. In fact, it's better not to know anything at all. For the E-Creative project, a rough diamond is more valuable than a highly cultivated precious stone. He'll hold on to this one for now.

Fucking matrix! The next candidate is Gem Matr'X, the company's star executive. Gatew felt obliged to put her on the shortlist. What a sacrifice, to have to get by without her! All the projects that are about to fail are sent to her and they all miraculously get back on track. She's never failed. Sus Cal'Vin calls her the "nullus defectus." Of course Sus can't stand her, she doesn't fit within her ever so limiting blueprints. It would be worth selecting her just to know which horrible defect would prevent her, surely, from overcoming the famous neuropsychological filter. But he's not interested in someone that old, no matter how many successes and good credentials she has to endorse her. He wants someone young, malleable.

Here's one, in her early twenties and conceived and delivered vaginally. Truly a rare breed if, as she claims, she doesn't come from an anti-techno commune. She must be using her origins to her advantage, otherwise she wouldn't have published them on the register. Gatew has fallen for it and maybe he's not far off: an atypical genesis can encourage an original personality, but other factors are necessary and, as far as he can tell, there is no evidence of this girl having any other merit.

"Let's see, two left... well done, Alph, first the women and then the men. You're such a gentleman!"

11:05 a.m. – I interpret that he's referring to the order of the candidates. I have passed them on in the order they were sent to me. I take note that, when a gentleman selects candidates, women have to come before men, and I leave open the possibility of generalizing this rule to other situations and recruiting. I deduct that being a gentleman is positive.

Demonstrating an exquisite dominance of 3D potentiality, a young Asian man has appeared on Alpha+'s monitor to explain to them what he's achieved and, above all, what he feels he is able to achieve. He gesticulates a lot and paces up and down and it occurs to the Doctor that the boy may be a little paranoid. He'll have to wait and see what Sus Cal'Vin has to say on the matter. The kid reminds him of those charismatic politicians they had two hundred years ago, who thought that just by expressing an idea the economic and arms-driven world system would be changed. How naive! The Doctor also believes in the strength of ideas, but they must always be kept secret, worked to exhaustion, polished and only at the very end brought into practice. To divulge them is to kill them off.

And this novice, look at him, all he needs to do is attach the blueprints of these suggestive mechanisms he proposes. Wireless transmutation. Happiness app. It's shockingly ingenuous. A person in a better position could take possession of his inventions without batting an eyelid. They'll end up featuring among the competition's products... Where are CraftER's censors? So many years fighting to impose the new regulations and they're not even applying them. Even Gatew's seen this and let it through! How can they all have forgotten the series of judicial battles they had to go through so that the ideas of their employees would legally belong to the company?

"Alpha+, I want a connection with Mr. Gatew immediately."

The manager appears on screen in a matter of seconds:

"Hello, Dr. Craft, I was waiting to hear from you. Which files should I forward to Dr. Cal'Vin?"

"None, for the moment. I've got Leo Mar'10 here in front of me. Why isn't his entry in the public repository censored?" He wants to hold back, it's been months since he handed this responsibility over to Gatew, but the anger is noticeable in his voice. "It's full of product ideas that the company could exploit."

"Yes, I know, that's why I forwarded his file to you. I knew this kid would be of interest." As silver-tongued as ever, isn't he, turning an insult into a compliment. "Indeed, you'll have the chance to meet him at the New Year's Eve convention, I've just appointed him to present one of his inventions."

"What? What kind of business model is this?" He needs to regain leadership of the company right now. "When have you ever seen a successful business that presents, as a great novelty, at its most important convention, a product that's available to everyone on the public network!"

"Calm down, Doctor, every one of these outbursts costs you four minutes of your life, you know that."

"Would you mind encrypting your pieces of advice? And then feed them to the worms."

"Sorry, I suppose I haven't explained myself properly: the device Mar'10 will be presenting has been developed within the company and, therefore, is not on the register. I guess you're worried about the ones that are on there. You can see potential in them, is that right?"

"Of course I'm worried about them, they belong to the company. What are they doing on the net?"

"Until now, that was the case, you're right, but a year ago an amendment was passed that allows non-retroactive contracts to be drawn up, letting people claim intellectual property they developed in the years excluded. No one takes advantage of it, because your wage is reduced proportionally. In fact, I'd never even read the amendment. But this Mar'10 came to us saying he wanted recourse to it in order to keep his private projects for himself; you must have noticed that he believes very strongly in his own possibilities for success. Considering he's a promising kid, something your interest is confirming to me, we decided to go ahead with the contract. Better to have him here than at the competition, right?"

He's irritating sometimes, but he was a wise choice for manager. The Doctor baptized him "the samurai of words" for his ability to take advantage of all the strengths of his adversary and turn them against him. He almost welcomed the nickname more than the promotion. He's a bit boring, that much is true, no outbursts, no raising his voice... but the Doctor knows that comes as part of the package. And, it seems, this is the case with Mar'10 as well. Great ideas seasoned with an explosive mix of immaturity and a whole lot of pretension. If he could take them apart and put them back together again as he wished, he could make some truly great men. It's got nothing to do with chromosomal selection, that's child's play compared to this. He's not interested in potential qualities that are often lost along the way; he wants realities, proven expertise ... the problem lies in how to isolate it. If he could extract the creative potential of Mar'10 and combine it with the wise, mature loyalty of Gatew ... that would be a cutting-edge ROB.

He's getting distracted again, and Alph's standing here with the next file ready, wasting time and letting him waste it. He frowns and looks at the robot angrily, in an attempt to make it understand that, in moments like this, when he's lost in thought, he should interrupt him.

11:43 a.m. – An indignant look after a long period of thought that I have respected. I interpret that he didn't like the last candidate. I close all the folders and I ask him if he requires another service or if I should leave.

"Fucking machine! Recover the previous state immediately."

11:44 a.m. – I make a note: when faced with a silence followed by an indignant look, it's better not to make interpretations, and instead, by default, maintain the current state, no matter how long it may last.

The name Miq 6'Smith rings a bell for the Doctor and, effectively, the man's face takes him back to the calamitous time he spent at the Department of Innovation, during which only this one deaf kid was capable of developing anything. Back then no one understood why Miq rejected a cochlear implant: "we waste too much time listening," he would say in sign language. And he was a trailblazer, considering the immense power that deaf culture has today, or the deaf species, as they like to call themselves. Even some hearing people, when advertising and noise pollution become unbearable, decide to plug their cochleas and join the tribe. A stroke of luck for CraftER. Idiosyncratic minorities are a great niche in the market: everything has to be adapted for them and that's our specialty. HandicapER, however, had to rethink their implant production.

It's strange that, in all these years, he's never heard anything about Miq. He avidly reads over his trajectory: projects and more projects, the majority developed alone and all pretty cutting-edge; he must be old by now: fifty-one years ... he would rule him out straightaway if he weren't the only candidate with experience in cognitive prostheses. Who knows, he might be the most reliable option.

By way of a summary, Alpha+ presents him a series of graphics and quantitative comparisons that give Gem Matr'X a slight advantage over Miq, closely followed by Leo Mar'10 and, farther back, the other three girls, who are practically even. He would have liked to be able to rescue Mar'10 from the last position and declare him the main candidate, against the prognostic, in order to feed the legend of him as a boss who is extravagant and capricious with the staff but still endowed with an indisputable gift for sniffing out talented young people. He's convinced that they have it, him the gift, and the boy the talent, although he decides to make Miq take on the neuropsychological filter as well ... and the girl with the Picasso-esque profile, in case it confirms that indeed she has no idea who Picasso is.

Once the relevant instructions have been given to Alpha+, the Doctor can go back to the battle. He presses the button and the riddle flows back into his brain. He imagines the monks, each one neatly tucked away in his cell, taking four paces up to a miniscule window, which maybe even has a couple of bars, with their hands clasped behind their backs before peering out, only to turn around and retrace their steps. Silent. Uncommunicative ... but united in a common meditation: what can they do to save the community. Perhaps he should dispense with all this scene setting and get to the point, the clock won't let him get away with it. Hug 4'Tune's proposals must always be thought of in computing terms: a series of identical processors that exchange data in a synchronized way, once a day; each one transmits whether or not he's unwell, without knowing it himself, that's the crux of it, and receives the information from all the others. At the heart of the question is lunchtime, that much is clear. If there are n sick people, each of them sees n-1 red marks, while the healthy people see n. The decision to commit suicide would be easy if they had been told how many sick people there were. For now, he's pretty lost.

Let's try looking at it from another angle: reduced examples usually give clues. Right, a community with only one monk ... leads to a stupid situation: he has to commit suicide because he must have contracted the disease but at the same time he doesn't need to commit suicide because he won't infect anyone. Making 4'Tune aware of the weak spot in his wording will give him a fractional advantage, he should have specified that in the convent there were at least two monks. He'll try it with two, then, only one of whom is sick, so that there can be a risk of contagion. The one who doesn't see the red mark commits suicide because it's obvious that he must be the sick one, and it works for any number of monks as long as there's only one infected. He's on the right track. Now what if there are two people afflicted?

An evil thought interferes with his thought process. If the message were a lie, no one would see any red marks, which would result in a collective suicide. What a clean way to eliminate whole communities in one fell swoop. Some of those anti-techno groups deserve it. Although, who can be sure that they're as credulous and altruistic as the monks. No one is like that these days, the closest thing to monks are probably the robots ... and, fucking hell, the last thing he needs is for them to auto-immolate. Goodbye, CraftER!

What a swine, he's losing valuable time that Hug 4'Tune will be taking advantage of to solve the riddle he gave him. At least he doesn't have to attribute that to decrepitude, his mind has always been undisciplined like that and now insists on reflecting on what would happen if the liar were one of the monks themselves, who, to make it all more believable, had painted a red mark on his forehead. After the first lunch, they would all expect him to commit suicide, but he wouldn't do it, of course, and, at the second lunch, each of the others would assume that he also had the mark and would kill himself. A lethal weapon, now operated from within. It's got some teeth, this idea. Hey, they would commit suicide on the second day, something's not right here, how could he miss that? If two people have the mark they'll commit suicide after the second lunch and if three have it, after the third. Eureka! A totally superfluous liar has switched on the light bulb, it couldn't have happened any other way. Who was talking about disciplining his mind? He hurriedly introduces the answer and raises his arms euphorically: prepare yourself, world, the great Craft has got plenty of fight left in him.

8

On Sunday, Silvana gets up at half past six, the same as every day. Even the warmth of Baltasar's body isn't enough to keep her in bed once she's awake, despite having made love last night with a virtuosity that she has never achieved with any other man, and knowing that the only worthy climax would be for him to open his eyes with her still in his arms. But no. She's felt different for days, free to transgress the most deeply rooted habits at the ComU, the ones that she helped to forge and that she still wholeheartedly believes in. Well, that's not really right. It's that she's become used to dedicating the early morning hours to getting lost in books and the emotions of two centuries ago, and she's found enjoyment in it. So much so that she's not prepared to give it all up.

She gets out of bed with more energy than ever, and, while she dresses herself head to toe in a tracksuit with her gaze fixed on the nakedness of her partner, a worrying thought sneaks in between the folds of her consciousness: bodies bore her. Standing stock-still with her arms only half covered, she repeats out loud, hypnotized: "Bodies bore me." That's definitely it, and it's been happening for a while, but until this precise moment she hasn't realized. The vague unease that comes over her while she gives massages might come from there, as well as the perplexity she experiences before the perfect musculature of some young people that no longer makes any impression on her. It bores her. That's exactly the right term. Not indifference, or rejection. The smooth and shiny skin would attract her if she hadn't seen it in so many interchangeable youths, and Balt… his naked figure, silhouetted against the white bed, was the last piece of the puzzle.

She squeezes her eyes shut to stop the runaway thoughts she doesn't recognize as her own. The body, senses… they're the only gateway into emotions that she knows. She can't disown them like this, they're the tools she needs to work. Her life and ideals are built on them. Oh, see, what's up with her these days? She nods her head decisively to expel these worrying thoughts while she zips her suit up to the top to protect her neck. These uncomfortable feelings are transitory, they'll pass.

Four quick strides later she is in the reading space and, before settling herself in the wrap-around armchair, she dedicates herself to the stretching and breathing rituals she performs every morning. She calls it revising the minuscules, the smallest muscles that always end up paying for the excess of the bigger ones, and only once she has confirmed that they are all awake and in the right place can she start her day. In the meantime, a flow of perception picks up the shy signals emitted from every corner of her body and puts them together to create a pleasant sense of well-being, of strength. And, surprisingly, she discovers that her body still interests her, which is lucky since it seems to be the only one that does.

She's at it again. Is this thought going to ruin her day? She opens the e-book and brings up the pending reading tree. In the biography branch, after consuming those of Mahatma Gandhi and Sigmund Freud, she abandoned Albert Einstein's as it was too technical, and instead got into that of his wife, Mileva Marić, written by a compatriot and contemporary: Desanka Trbuhović-Gjurić. The name blinks in red on a terminal branch of the tree. It's not the experiences of Albert, nor even those of Mileva that interest her, but those of Desanka. A completely unknown woman. Strongly affected by the personality of the genius's wife, she dedicated herself to paying homage to her and making sure she went down in history. A curious expression, that is still understood but without anyone ever making use of it… there's no such thing as history anymore, only a flux that records itself, collectively, without critical individuals, or, if there are, we lack the ability to recognize them. Maybe there are still Einsteins, but the Milevas and Desankas have died out and, with them, the possibility of going down in history has become history. She smiles. That slogan sounds more like Baltasar's than hers, and the fleeting reference to his body silhouetted against the bed impedes her thought process again. Boredom when faced with bodies, had those women ever felt that? Or is it a modern feeling, as impossible to transfer through time as Mileva's self-abnegation or Desanka's admiration of another? Self-sacrifice as a way to highlight another person's merits is a concept that no one understands nowadays, not even her, and she has no interest in recovering it, but reverence before an almost superhuman quality is an essential accompaniment if the former is to survive… without fans there are no heroes. Without the ability to differentiate or the desire to emulate, all direction is lost, and the boomerang becomes inevitable.

She lifts her gaze from the text, surprised at how she's linked everything together. Recovering admiration in this way will become an important step toward the milestone they're all after. Just the argument she needs to convince the Ideological Committee to dedicate more resources to her sentimentality research. Until now neither Baltasar nor Seb have fully supported it, and they're big fish. Neither of them has told her so openly, but she's certainly noticed it. And it's logical. A whole life trying to bring people closer together, physically, fighting against the devices that distance us: no more virtual work, no more electronic contact… and not only in social and work-related fields, but also in the most intimate circles. That's where the ComU came from, and that leitmotif that Baltasar used to stir up his followers at meetings, "we must touch skin," he shouted, and with the same words, whispered into her ear, he ignited her like a torch. She squeezes her eyes closed to undo the spell that seems to be possessing her today. It's been years since she's thought about those words, so badly timed now, but at the same time so pertinent. The subconscious, Freud would say, is always in the shadows and well aware of the games being played on the surface. She covers her face with her hands in an attempt to contain all these uncontrollable associations and organize her ideas. Her mind is working like the e-book now, with boughs, branches and twigs, trains of thought aborted here and there that she would like to take up again, though she lacks the tools.

When she opens her eyes, aside from Desanka, she sees other flashing markers: Zweig in the memories section, Freud-Abraham in letters, Flaubert in romantic novels, Solomon in psychology texts… she'll do the same thing, take note of the pending threads so she can return to them when she feels like it. First, she needs to focus on admiration as a key tool for stopping the boomerang. Second, despite physical distance being important to this affect, the net result seems to be a marked proximity, a different kind of contact, according to the descriptions of the biographers and psychologists she has consulted. It's not a million miles away from the aims of the ComU, then. Third, and here is where it's harder for her, until she'd remembered that slogan about the skin she hadn't made the connection: is her interest in this distant affection what has provoked the disinterest she feels toward bodies? If that were the case, the committee would have a strong reason to object, to avoid it spreading like an epidemic. And if what's happening is the opposite? That, without realizing, it's the malaise that's brought her to these men and women who are disembodied but loaded with more powerful affections than those found nowadays?

She would like to think that today's unease has nothing to do with this months-long pursuit, that tomorrow she'll wake up with her passion for bodies and skin renewed, and that she'll be able to go back to normal, to the life that's always fulfilled her, while at the same time continuing with her research and books... but the doubt is too strong, it's become an obsession. She goes to stand up and give up on her reading, just as Baltasar enters the room. Clumsily, but tenderly, he sweeps back her hair to kiss her on the forehead, before fixing her with an interrogative but still sleepy gaze.

"I know, I know," she says, "but I've reached a really interesting part, and you were still asleep..."

"Those ancient leaders again?" He kneels down to see what she's reading. "I'll have no choice but to challenge them to a duel. So many distinguished names, it could be a pantheon. Who's this Desanka fellow?"

"She's a woman."

"Wow, you really are making it hard for me, I'll never be able to fight her. How do you challenge a female rival?"

"Come on, stop with the foolishness. What time do we need to leave?"

"Don't change the subject. Tell me what's so interesting about this Desanka person."

"Read it yourself." She turns the e-book showing the paragraphs she's highlighted round to face him.

> ... Mileva's personality began to interest me many years ago; the depth of her thought and emotions moved me ... I've started writing to do justice to her thirst for knowledge and her search for the highest values ... I cannot allow Mileva to be forgotten while her husband receives all the public recognition.

Baltasar, surprised, looks up at her:

"Now you intend to dig up the feminist movement?"

"Negative." Along with the sing-song tone, Silvana lets out the first laugh of the day. Suddenly she feels much better.

"Who was the greedy husband?"

"Albert Einstein."

"My God! Now you're confusing me. Since when have you been interested in science? And ... what point is the biography trying to make, that the theory of relativity was invented by his wife?"

"Not quite that much, but it seems she played a decisive role. She wrote all the mathematical formulas. Yeah, yeah, don't make that face, Einstein recognized her part when he won the Nobel Prize. They had already divorced and he had remarried, he sent her the prize money in recognition."

As soon as she's said it she realizes that everything's fallen into place: the admired genius becomes the admirer ... and this second role, in her eyes, magnifies his greatness. It's obvious that Albert and Desanka revered different qualities in Mileva, each one appreciated what they saw according to their own filter, there aren't enough filters nowadays ...

"I see the subject excites you." Baltasar, still kneeling, was watching her while she thought things over. "And ... it's giving me bad vibes!" All of a sudden he stands up, and runs his hands over her body, kissing her on the neck, the ears, his tongue inside them ...

And Silvana manages to note down the last thread, "filter," before melting into the liquid tingling feeling that's heading for her eardrum. No traces of thought, no inopportune flashes of a body silhouetted against a white bed. Only his tongue and those hands that are always the right temperature.

9

It was the last recommendation the psychologist had made before Celia was due to leave the clinic, and Lu was paying less and less attention to what she had to say because she didn't even know where to start. How could she organize a party like the ones they had a hundred years ago? With other children, games, and sweets, if she didn't have any kids at hands, games had never interested her and just reading the definition of "sweets" in the glossary had made her nauseous? She would have buried that piece of advice forever if it hadn't been for a more experienced adoptive mother she'd met at the gym the other day who gave her the key, she had to get in contact with the ComU. They were specialists in all kinds of archaic celebrations and, for a reasonable price, they would take care of even the smallest details.

And she was right. When the day came, she only had to open the door and, in no time at all, her house was no longer home and had become instead a party venue from the turn of the millennium. Her beautiful hanging semispherical chairs had been pulled right up to the ceiling, along with the rest of her suspended furniture, and had been replaced by soft, shapeless seats, which had been spread across the room. There was not one screen or control panel on show, as they'd been hidden behind an elastic material she had never seen before that simulates openings onto an external landscape darkened by vegetation. The living space had been so luminous before, with its unique yellow walls, but now, to break up the shadows, they've had to add points of light here and there, and hang long, colored paper chains to brighten the atmosphere. Of course they'd assured her that, once the party was finished, they would only have to peel off the elastic cover, a kind of indoor bubble, and the house would be good as new, with no breakages or mess.

She hasn't had to concern herself with Celia's party outfit either, or with inviting the three boys and two girls who arrived punctually and at the same time. At first the black boy, who is extremely tall, seemed to her to be much older, and the Asian-looking girl much younger, but the party animator made it clear that they were all the same age and that these kids would be her daughter's classmates.

Her daughter. Her heart still skips a beat every time she hears it. They put too much emphasis on the word, as if they were conferring a lot of respect onto it, when she would rather see it as just another label. Luckily the little girl didn't insist on calling her mother. She says she already has a mother, and that she talks to her from time to time. Poor thing, she really is lost. Maybe it's true that since she's pretty grown-up it's more difficult for her to adapt, just like they warned her at the clinic. Last night the girl called her "auntie" and she had a hard time understanding that it was meant to be an affectionate name. According to Celia it's the name given to a grandmother's daughters, but nowadays very few grandchildren stay in contact with their grandparents, and fewer still with the daughters who don't happen to be their own mothers. How complicated, and also, of course, figurative.

She's getting bored because they've stuck her in the corner so the party could go on without her having to participate. What they're doing is of no interest to her. Only the little blonde girl, Xis, with a blue ribbon in her hair and matching pointed shoes, is a pleasure to watch. She realizes that Celia is looking over at her. Maybe she's been watching her all this time while she thought she was hidden. It's a bit overwhelming to find that Celia's always monitoring her, as if nothing could really keep her occupied. She would prefer it if, like the others, she only had eyes for the young animator, who is arranging five chairs in the middle of the room and proposing that they play a very simple game, which she says is for little children, to break the ice and decide who will sit where. She gives some brief instructions and the kids start moving about: two spin around in circles, a third runs off and sits down, and Xis takes a chair and drags it over to the side of the room, while the Asian girl, having seen what she's doing, drags another over to the opposite side. Celia, who had started running, has stopped and is staring at the animator, and by the look on her face, Lu realizes she doesn't understand what's going on. She was distracted watching her and now they have to start again.

But the boy who is sitting down refuses to get up and Xis doesn't want to give up her seat either. They already have a place, now the others have to play. It's logical, but the animator insists; she must be trying to make them understand that it's Celia's party so if she makes a mistake they have to pretend nothing happened, but she's struggling. The boy wants to get his own way no matter what, and in the end they have to let him connect to his games channel. Xis amuses herself by sculpting the soft seat into different shapes. The other four, having received new instructions, start the game again, but now with three seats.

They run in a circle around the chairs and, oh, the animator has screamed. Lu knew they'd end up getting hurt. Celia has bumped into the black kid, rebounded off him and ended up sitting down. He's giving her a very angry look and it seems to Lu that she must intervene.

"Is my little princess hurt?" She moves forward with her arms outstretched so she'll feel safer in unknown territory.

"No, why?" Celia, confused, avoids this outpouring of warmth.

"That jerk has knocked you over. Has no one explained to him that you can't touch girls?"

"But it was my fault, I pushed him when I was sitting down. And he didn't touch me," she blushes, "really he just moved me out of the way."

The boy still looks offended.

"Calm down everyone, one moment." The animator takes a step forward in order to compensate for Lu's meddling. "You were all doing very well. That's the idea, to sit down as quickly as you can when I make the signal. Someone has to be eliminated every time, son, that's the game."

"She cheated. I have the right to keep playing."

"I'm sure you know some more peaceful games, right?" Lu is using the most authoritative look and tone she is capable of.

"Of course, ma'am, after we've had a snack we'll play twenty questions and pin the tail on the donkey, and to finish we'll do karaoke. But children have trouble learning to wait their turn, and physical games can help them all get involved, since they can all be protagonists at the same time."

"Oh, no, that simply can't be, the only protagonist here is my little girl. It's her party. You should have started off by asking her what she wants to play."

Suddenly, Celia becomes aware that she's the center of attention, even though no one is looking at her. She feels she has a responsibility to make a choice that pleases everyone.

"Could we dance? We're three boys and three girls, and with the karaoke music ..."

"No, darling, none of them know how." The animator turns and addresses Lu: "At the ComU we do courses; if she wants, for a future party we could teach them ..."

"But we don't have to do it properly, holding each other and all that." Celia isn't prepared to give up on an idea that seems acceptable to her. "We can put music on and everyone can move around however they want. That's a physical game where everyone takes part at the same time ... and there are no winners or losers."

There's a bit of back and forth between Lu and the animator before they come to an agreement about what they should do:

"Right, guys, we'll come back to this later. My colleague has just told me the food is ready."

***

Hi, Mom, can you see me? I have so many things to tell you about today that I don't know if they'll fit through such a small hole. Lu organized a surprise party for me. Fifteen days without seeing anyone and today, all of a sudden, five children and two animators. The one who seemed nicest is Xis. First she was sulking in the corner, but at snack time we started talking and she told me loads of stuff about the school I'll be going to. That really will be an adventure, just like you said. Apparently there are no desks, not even a classroom. They go from one place to another whenever they feel like it, but they're separated by level, of course. The teacher doesn't order them around, he only advises them. Imagine that! She says there are lots of gadgets and they're really fun to use. You don't have to study, only experiment and answer questions. It reminds me of that museum we went to with Dad, do you remember? You could touch everything and he loved showing me where to find the ribs, the liver, the stomach ... he said "I'll introduce you to my colleagues" and it made us laugh. He made me pick up that huge stone too, if it had fallen on me it would have squashed me flat, but thanks to those pulleys it hardly weighed anything at all. Some of the miracles of physics are exhibited here, they told us, but what we wanted to know was if all miracles can be explained by physics. I think about these things a lot now that I'm here, so far away. Maybe I'll find out how to turn back time and be with you again.

For a moment she is struck dumb, with the ring in her hand and her eyes welling up. She doesn't want to cry. Today she's challenged herself to be strong of heart, and for her mother to see her looking happy at least once. She must make the most of the nice things she has to tell her.

I haven't told you the best part yet. At the end of the party Lu gave me my very own robot just for me. Yes, a robot, I know it will be difficult for you to imagine. So you have an idea, it's like the ones from Star Wars, but, along with legs, it has four wheels for when it wants to move quickly, and it doesn't have a face. Well, it has a kind of head with no nose, mouth or ears, it just has two cameras, and a screen embedded in its chest. It's called ROBbie. I'll have to learn to use it, even though it does a lot of things on its own already. It will go with me everywhere; to start with, we'll go to school together tomorrow. I didn't know, but everyone has their own robot here; it's like us having a wallet or a calendar, but much more sophisticated, because it has a large memory and can solve problems for you. Lu's is called ROBul, and it's been hidden all this time so as not to scare me. I don't understand why. I've been more shocked by the kids, and even some things Lu does, than by ROBbie. For the robot, everything follows a series of rules, it'll never surprise me with anything inappropriate. I can't wait to test it. For now, I've asked it to find me songs by The Shins and Arcade Fire.

Oh, Mom, this all seems like a play where I'm the only living person. The others are like cardboard, or stone ... or mechanical, like ROBbie. What am I doing here, if no one cares about me and I'm having a bad time? There are moments when I feel so horribly alone that I'm dying of fear. Then I give up hope and I feel like I could kill myself. If I don't, it's for you and Dad. Maybe I haven't understood where I am, I tell myself, and I try to calm myself down. Who knows, it might just be a bad dream and you could appear at any moment. Sometimes dreams are very real. Why have you chosen this for me, without asking? I don't understand, Mom, you always told me everything!

But where has this come from? She's let go again without realizing. She had such good intentions but at the first test...She has to get this out of her head, or she'll feel really depressed later and she won't be able to stop the bad thoughts. Luckily she has the ring, at least that's real. She points it upward again.

Forget what I just said, Mom, I don't know what came over me, it's not so bad, I mean it. With you watching over me and giving me strength the whole time, I'm sure I'll get through it. And maybe in the end I'll understand why you sent me here. Lots and lots of kisses for you and Dad.

II

NEW YEAR'S EVES

10

For CraftER's twenty-fifth anniversary, the company has decided to convert the traditional New Year's convention into an unprecedented event. They've bought a hundred generators to shore up the supply of light for just this one night. In addition to the usual sensorial stands, wrap-around screens and personalized sound systems, they've installed a display of the lasers and speakers used at previous conventions in order to highlight the durability of the company over the years, and, to take that idea even further, its timelessness. "We're the best, we have no rivals," is the triumphant message that is disseminated though all subliminal pathways available to their employees who, dressed up and smiling, have turned up for this unmissable occasion.

And Leo feels like the best of the best. Mr. Gatew has chosen him, and only him, to represent the centro-european headquarters at "The Product 2111" contest, the star forum of the night. He knew his sensory transmutation booth would impress Gatew, but he hadn't imagined in his wildest dreams he would have the opportunity to show it off in front of the company's top brass, Dr. Craft included. He paces up and down, nervous, euphoric and a little uncomfortable in the official dress that identifies him as a contestant. He has a couple of hours to enjoy the various ambiences before his performance. While on the main stage they give out awards and pay homage to those who have worked at the company for twenty years and ten years, respectively, he amuses himself at the Disasters stand, where they make fun of their rivals' worst inventions. At the previous two conventions he got his best ideas from this stand, and, little by little, he has convinced himself that one small error can be much more productive than thousands of successes. What can you learn from a finished design? Nothing. On the other hand, a failed attempt is always a challenge, a desire to get to who knows where, a bombardment of suggestions. And, precisely, a missile has just hit him. He's not sure what drew him to this realistic mechanical baby, if it's the grotesque expression, the diapers it doesn't need or the fact that it bears the logo of Bet's company. First they brought out those practical little dogs that didn't need to poop or pee, and then they started mimicking wilder and wilder animals, until they got to man. What woman could resist the charm of a baby that smiles when she coos at it, that she can cuddle at will while watching her favorite program, that recognizes her voice and crawls along behind her, flattering her with sweet noises? And, best of all, that can be turned off and shut in the cupboard when it gets whiney and tearful? Well no sir, the product didn't take off, almost certainly because it's too much like the real thing, déjà vu.

He calls ROBco, as he's had to leave him in the wardROBe this year, and tells him to phone Bet.

It's an inconvenience not having his robot with him, in past years it commented on events, they recorded anything they fancied and he could instantly get in contact with this or that person, but this year the higher-ups had been strict: they wanted all the employees to bring their partner, or a friend, or a neighbor, as if they cared; the real reason is that they want to extend the celebration to new potential clients and reduce the competition's audience. Mr. Gatew had insisted on his bringing Bet, but she refused as she risked being fired. And the truth is, like him, most people are wandering around on their own, missing the old conventions where they had their ROBs for company.

Although he was expecting the incoming communication, the soprano squeal makes him jump:

"Hey, Leo, how's it going? Have they announced the final figures yet?" Bet always gets straight to the point.

"No, not that I know of, of course I'm not on the main stage and I haven't tuned into it."

"You're hopeless. You have to put yourself out there, Leonix, today is your big day. We've gone up 23.92% and we're third by year of creation and seventh in the technology sector."

"I'm pleased to hear it, I can tell you're happy."

"And that's without counting the points that the Arbitration Council might give us ... we could even go up to fifth place ... can you imagine? Fifth place, Leonix."

"Great, great. ROBco is warning me our president is about to make a speech, I want to give him my full attention, you understand that, right? I have to hang up, I'll call you back when he's finished."

Dr. Craft, a man dressed in black who talks with his eyebrows, dominates the stage, multiplied by all the wrap-around screens, virtual ambiences and liquid mirrors surrounding Leo, who decides to sit in an octopus massager to wait for the New Year's message. A tentacle repositions his vertebrae one by one all the way up to his neck, two more wrap around his legs, and another couple gently rub his elbows and wrists, but none of it reaches his brain, which is totally sucked in by what he's seeing and hearing.

"CraftER employees, we've done it, we're number one"—he stretches his arms up in the air in a victorious gesture—"in our category ... and over all!" A great cheering is heard and images of an ecstatic crowd are superimposed over that of Dr. Craft. "Our products have no rival. We achieve the best results in the majority of categories and the competition doesn't know how to stop us. Throughout this year they've tried everything: snatching our suppliers away from us, discrediting our distributers, disparaging our aftersales service, they've even tried to tempt our best staff away with big offers. But no one wants to leave CraftER, and it's a matter of logic: we're the only company that guarantees your future." During his studied pause the monitors are again filled with superimposed images and unintelligible cries of acclaim. "And next year will be even better, I promise. We have projects that will surprise even the most daring among you, and I know you're out there: young people that believe so strongly in their own inventions that they only accept non-retroactive contracts."

Leo feels a strange prickling sensation: it seems like Dr. Craft is talking to him ... but that's impossible, he doesn't even know him. It must have something to do with the communication assessors, ever better at modulating discourse so that each person receives it as if it were meant only for them: a little allusion and everything is given a personal, exclusive edge. The one and only acoustic source with as many meanings as listeners.

"Ingenuous young people: we'll leave you dumb struck and, when you come running to me pleading for a contract extension I'll say ..."

Is he threatening him?

"Welcome to CraftER!, because I like people with ambition and it makes me proud that you identify with the company."

Nice start, president. Leo had never heard him speak and he found him to be wittier than he expected. When a round of live connections is announced, he gets out of the octopus, which has given his muscles a new lease on life, and heads for the area reserved for contestants in "The Product 2111." In a few short minutes his sensory booth is ready for action and, with growing anxiety, he prepares himself to sit through the presentations of his opponents. It would be great for him to win! Then he could really develop his idea right through to the end, and be sure of it working. He just needs resources ... lots of resources, and with CraftER behind him, he would be guaranteed to have them. A dream, having all the company's machinery at his fingertips. No wonder the Doctor is so satisfied—he's teased that incredible projects are in store. Who knows what he's hinting at. But how can he let his mind run away with him now? He must concentrate on the presentation and convince the president that his product is the right one, the one that will bring the most money and prestige to the company. He has a good chance: the man himself has admitted that he has a weakness for young people with ambition ... and he's the youngest participant.

When Mr. Gatew announces that the audience will decide the length of the presentations using the sudden death system whereby they end when two-thirds of the audience members have disconnected, Leo notices the people around him for the first time, men and women in passive mode, with their lives rented out to CraftER and with the sole ambition of staying there until the end of their days. At least that's how he sees them. Right away he asks himself what he needs to do to make sure more than a third of these puppets don't take their attention off him. But before he finds the answer, the speaker specifies that choosing the winner will be the prerogative of the president. He doesn't need to worry about other people then.

A minute passes before he realizes his error. If they don't allow him the necessary time to explain himself, he won't be able to convince the

Doctor. They've thought of everything, these machi'vels, you have to get through the filter of the masses in order to be heard by the only audience member who matters. Now, during his presentation, he'll have to entertain the auditorium with gags and immediately solvable mysteries while conveying the genius of his invention to the judge. And he'll have to give up on his dream of the Doctor trying the booth, such a personal tool that it would turn everyone else into nothing but spectators, envious and frustrated. It would be suicide. Neither the Doctor nor anyone else will try it out, everyone will see how it works together. From the presentation tree he's prepared, he chooses the one that contains the sequence of Mr. Gatew standing in for Michael Jordan, pulling off unbelievable dribbles, leaps and shots. That'll keep them hooked for a while, halfway between surprise and mockery, only to succumb to the explosive enthusiasm of the manager as he left the booth after his transhuman experience. Guaranteed fun for the masses and the best possible recommendation for the Doctor from his right-hand man. Everyone will be desperate to get a taste of his invention, and he'll be speculating about the possibility of choosing someone to win some precious time that he will fill with technical virtuosity and, above all, flashes of the limitless human perception that implementing his wireless idea could impart.

With each new contestant that steps up to the stage, Leo feels more confident. The one who took the longest to surpass the two-thirds threshold was only able to release four motes of his sensitive space dust. The idea of extending oneself in floating sensors capable of getting into unexplored corners where only dust can go seemed suggestive enough, but they didn't even let him set out its possible applications. The masses are demanding today, and as soon as you let your guard down they take their support away without a second thought. He must be careful.

The man who's up before him looks like a descendent of sea creatures, and, before he even manages to differentiate his metarobot from a standard reconfigurable robot, he's already been forced off the stage.

It's his turn.

Not wasting a second, he plants himself next to Mr. Gatew and turns him into a copromoter of the product that he's developed at the centro-european headquarters. He was the first to believe in the product and try it out for himself, he fully deserved to be on stage. The manager smiles proudly, and Leo becomes aware of the force of the Doctor's gaze weighing down on him. For a few moments he has his blessing. And he must take advantage of it. He walks in and out of the booth, he shouts, he gesticulates wildly as he plays with the 3D images, enriching them with comical scenes: standing up to the giant players and gushing with admiration for Gatew the infiltrator, who, astonished, doesn't know where to look. He possesses a prodigious energy; *in vivo* he electrifies the promotional film even further, Dr. Craft thinks, with a malicious smile that puts Leo off his stride. As soon as he stops moving he sees the attention meter plummet ten points in one go. He can't allow himself even a glimpse of that face and, concentrating on his self-control, he throws himself back into the presentation.

He manages to survive twelve minutes and fifteen seconds, quite a feat. When he leaves the stage, he takes the liberty of scrutinizing the Doctor again and, beneath his threatening eyebrows, a hard gaze meets his own and follows him until he's out of his line of sight. It's like being touched with the blade of a sword, Leo doesn't know if he's been mortally wounded or chosen for the highest quest. The few minutes he has to wait for the verdict are unbearable.

Mr. Gatew helps him pass a few minutes by coming over out of the blue to tell him that the president wants to see him. All of a sudden his heart leaps into his throat, with such force that he can't even articulate the question that's suffocating him. Has he won? As he walks along behind the manager, the idea starts to sink in and he practically floats along on a cloud of euphoria. Bet's face when he tells her... She was always reprimanding him for wasting time on his "inventions," as she called them, discouraging him saying that he would never manage to move up in the company by working on private projects... and now, all of a sudden, he's at the top: CraftER will sponsor his dream of creating wireless transmutation. All the work has paid off.

He doesn't know how they got there, or how long it took, but now he's standing in front of the unmistakable Alpha+. A watershed in robotics, he would love to experiment with it; what a shame that the one day he has him within reach, it's not the right time.

12:25 a.m. –"Dr. Craft has proposed a brief personal meeting, during which you will be unreachable, and will not be recorded in any way, do you accept?"

How could he refuse the opportunity of a lifetime? He'd even agree to go in there naked. After undergoing a short interrogation and cutting off his connection with ROBco, he is led into a small, luminous space, where he finds himself alone. Four walls and two transparent armchairs, nothing more. Absolute isolation in such a small space makes him a bit anxious. He doesn't like power games.

The Doctor's entrance has more of an impact than he'd expected. It's nothing to do with his physical appearance, which he was already familiar with, nor with his gait being less unstable than expected. It's his commanding gaze that seeks to rob him of his will, and against which he feels defenseless. No one has ever looked at him this way before.

"Do you know why I've summoned you under such exceptional circumstances?" He sits down without taking his eyes off Leo, who does exactly the same thing.

"I'm not entirely sure." Finding that he can speak helps him brush off his fear.

"You got the largest audience, do you think you've won?"

Leo doubts that triumphalism will get him anywhere:

"It doesn't just depend on the audience. It's you I had to convince. Did I manage it?" He clings to the hypothesis that they have the same taste in technology.

"Yes and no. I've chosen you, but not your product."

"I haven't won." The dejection returns him to his old impulsive self.

"I said I've chosen you, don't you understand that?" His eyebrows form such a pointed V it's like Leo can see an arrow coming. "I'm offering you a proper project, not this product 2111 bullshit. You will build a creativity prosthesis, it's just what it sounds like."

"A prosthesis? That's more HandicapER's business than a CraftER thing, and all this secrecy ... do I have to develop it outside the company?"

"I like to see your mind running away with you, but you're not quite right. It involves undertaking research that must be carried out under strict confidentiality. Apart from you and I, only Mr. Gatew will be in the know. And the prosthesis ... I call it that, but it doesn't have much in common with what the competition is making. It won't augment a physical function but a mental one. It will be pure software, let's put it like that. A program that will constantly challenge its PROP to spur his inventiveness and to avoid him falling into routine patterns. A kind of critical consciousness."

"Sounds attractive, but... what about the projects I'm working on now?"

"We'll assign them to someone else."

"Is everything already set in stone? What happens if I say no?"

"It's up to you. With CraftER as your enemy, you can forget about your ambitious idea of wireless transmutation."

Once again the sword is hanging over him, but Leo takes the risk:

"Do you think it's possible?"

"After the prosthesis, who knows. It could be the product of 2112 or 2113..."

"And why not 2111? Who won?"

"The zero maintenance robot."

"The one that repairs itself with waste materials? But no one was interested in it, they didn't even give him two minutes." Leo is utterly bewildered. "Do you really think his is a better product than mine?"

"Son, you don't understand a thing. I don't care about the products, all I want is for you to be available to work on the prosthesis."

"So that's why you put me out of the running in the competition, despite having held out longer than anyone else. It's not fair!"

"What is fair is only decided by those that have the power to decide; anything else is just bullshit."

"My device has potential and you know it." He's outraged that they've cheated him like this, so much effort had gone into the presentation and all for nothing. Bet was right.

"Look, Leo, that's your name, right? I don't give a shit about your product: what I'd like to transmute myself into doesn't exist yet." He pauses, as if he were thinking something over. "But, with the prosthesis, we'll be one step closer to it. I have made you the offer every CraftER employee dreams of; you should be over the moon."

"Yeah? And what do I get out of all this?" He realizes he's being too forward and corrects himself. "I mean, how will my job change?"

"To start with we'll update your ROB to the Alpha+ model, with the learning module, the neuroaccelerator... well, what more do I need to say? You'll be flying in both body and mind."

"Will I have access to it at home or only at work?"

"You know full well that's against the rules."

"But this project is extraordinary..."

"You're right, the rules are there for me to break in cases like this." He tunes himself to the kid's coolness. "You will have access to the ROB twenty-four hours a day. I only ask for discretion. And, if I'm satisfied with the prosthesis, you will earn a full wage without having to sign the retroactive contract."

"Oh, you heard about that."

"Yes, and for heaven's sake do me the fucking favor of taking down everything you have on the public register. If you just work like I hope you will, we'll talk about it soon. Now go and sign the agreement, do what Mr. Gatew tells you."

"What agreement? I haven't said yes yet."

"Fucking ego, you're more useless than the scrap metal dummy." He stands up and turns away from Leo. "Tell Mr. Gatew to send the next one in."

"I accept, I accept!" It burst out as an imploring cry.

"Rule number one, boy: with me, never risk anything more than you're willing to defend to the death. Any more funny business and you're out of CraftER, understood?"

The arrow hits home this time, despite him not being able to see the eyebrows, and, as he returns to the real world of the convention and the farcical contest, the only thing he wants is to face up to the challenge ahead of him, be worthy of it, and be strong enough not to succumb to the threat that's still ringing in his ears.

11

Silvana, with Baltasar's arm around her shoulders, joins the flow of people heading for the great hall of the ComU, but soon has to detach herself in order to greet a series of acquaintances. Hugs, kisses, handshakes and best wishes for the future, identical to those exchanged last year, but today she sees them as worn-out, expired, incapable of making her feel warmth or happiness. This feeling of indifference makes her shudder as she thinks that, outside her professional life as well as within it, skin and physical contact are no longer a stimulus. Maybe it's because she's obsessed. She shakes her head to dispel her worries and, taking Balt's arm, goes over to greet the organizers of this year's celebration.

Yeong and Sun, two long-established members of the ComU, welcome them at the door, wearing traditional Korean dress, and, between bows and curtsies, invite them to adorn themselves for the occasion. They're advised that the appearance and attitude adopted on this day establish a theme for the rest of the year, so it's best to choose bright colors, which will assure a bright future. From the display of colored silks, where eye-catching and warm colors predominate, Silvana chooses a red dress, the collar and cuffs of which end in a black bow. In the meantime, Baltasar observes her with a lustful look on his face, from within some shiny blue overalls and a pointy hat that makes him appear even taller. The hosts approve their choices, while at the same time insisting that what they do today will set a precedent: in the old days people abstained from telling their children off so they wouldn't have to put up with their crying throughout the year to come.

When they pass through the curtain into the great hall, they're left speechless; the space has been transformed beyond recognition, making it impossible to distinguish its dimensions or find one's way around in a place where they have spent so much time at numerous meetings and events. Ramps, folding screens decorated with stylized figures, gardens of vegetation, long, vertical strips of red paper covered in characters written in black ink, dragons made of bamboo, silk and cardboard, and lanterns hanging from the ceiling here and there that emit a yellowish light into the incense-heavy air. They really are in another world. And all without using any kind of virtual technology, as is stipulated. The recovery of traditions to mark the change of the year has really taken off, and each year surpasses the last, raising the bar even higher.

"Do you remember the Roman one?" Baltasar whispers into her ear, and she smiles, remembering how they had squeezed into a Janus costume, the two-headed God who gave his name to the month of January because it looks backward and forward at the same time. He wraps his arm around her waist taking care not to crease her dress, and looks her straight in the eye before kissing her sweetly. "I'm guaranteeing our future"—and, stepping back a little in order to make a point of the wall of clothing that separates them, murmers—"it was easier when we were dressed as Janus, though."

These are the last moments they spend alone during the whole celebration. Right away, some friends pull them over to the area where they will soon pay homage to their ancestors. There are a series of dishes on a high table, organized according to a color scale that forms a somewhat peculiar rainbow. On the left, red fruit, pieces of meat, tomatoes, dates, nuts, vegetables and soups, and on the far right, rice, fish, and a whitish liquid. Some Asian youngsters, who must be visiting, as they've never seen them before, encourage the attendees to write the names of their dead family members on a large wall, pointing out that the top part is for great-great grandparents, and the bottom one for parents.

The only place there is any density of names is at the bottom, and Silvana regrets not being able to contribute to the top part. Her mother never spoke to her about her origins, and, actually, she had never shown an interest herself, it's only now that she feels attracted to the past. A Korean man with a long, white beard stands before the altar and bows twice, deeply, his forehead almost touching the floor. On either side of him, two elderly acolytes indicate to the audience that they should be quiet and bow their heads. The officiator pours liquor into a glass and holds it up toward the wall, a gesture that is followed by other Korean members hurrying to place spoons and chopsticks into different bowls of food while they perform a prayer, during which they are spontaneously joined by other partygoers. Someone behind Silvana whispers that it's time to choose the food their ancestors most enjoyed when they were alive and to offer it to them as a sign of respect.

Some children shouting and murmurs of conversation can be heard from afar, which distracts some of the attendees, but not Silvana, who is concentrating on reliving what people must have felt when this ceremony was performed in the past. The strange mood, the scent of incense and the silence set off her imagination, and she shudders upon noticing an unknown spark inside of her. Baltasar, ever attentive, asks her if she feels alright, and she says yes, very quietly, so as not to interrupt any process that may have started up within her. She knows that this indiscriminate respect for one's ancestors is not what she's looking for, there are no filters here ... but it's similar. And, who knows, maybe with the right stimuli she will be able to uncover the underlying emotion, just as laughter brings happiness and not the other way around.

As the chants fade away, so does the movement of people heading toward the altar with their offerings, and it finishes when the acolytes perform two complete bows and, amid a captivating quiet and stillness, they prostrate themselves at the old bearded man's feet. Silvana feels a shiver down her spine, she's moved by a bodily configuration she has never seen nor imagined, that is capable of making her hair stand on end without any form of physical contact. This time Baltasar's interrogatory look is not met with a response.

Once the dead have been honored, the living get their turn, eldest first. The officiator is first to receive a glass and a bowl from one of the acolytes, while the other encourages each of the attendees to offer food and drink to someone older than them, and then to accept the offering of someone younger. A lot of worried faces look around them in an attempt to estimate the ages of the others, and only find equally confused looks, until some festive music breaks the formality of the moment, and everyone starts to move around in what is now a totally relaxed environment.

Baltasar and Silvana are still pretty far away from the table when Sebastian comes over to them holding a bowl of broth with little white squares floating in it, as well as meat, vegetables and a thinly sliced fried egg, the traditional *ttokkuk,* as he informs them. He's also carrying two little sticks and a large glass of liquor in the other hand.

"Sorry, young lady, but it would be rude to offer it to you," he says, lightly brushing her cheek with his lips, before handing the bowl and glass to Baltasar.

"Come on, get into the swing of the party, you fool!" Silvana exclaims, in an unconvincingly jokey tone. "To call someone old today isn't an insult, quite the contrary, it's an honor; haven't you caught on yet?"

"Excuse me, ma'am." He takes both of her hands and kisses them. "Seeing you in such youthful dress made me forget that I had the greatest great-great grandmother before me." He takes the bowl from Baltasar and offers it to her."

"Did it just come out like that or did you plan the bit about the greatest great-great grandmothers ..."

Sebastian is about to reply but Baltasar cuts him off:

"Now that I've got you all entertained with your word games"—he puts his arms around their shoulders and gently pushes them toward a corner where they can sit down and eat—"I'll take advantage and contribute to the ceremony while digging for useful information. You don't mind, do you?" he adds, smiling at Silvana before kissing her, with his arms still around both of them.

They're silent for a few seconds, observing how several people come up to say hello to him as he walks toward the altar.

"Maybe I should go and complete the ritual too, don't you think?"

Sebastian takes her arm and makes her sit back down.

"Forget it, darling, there's no need to be so strict about these things. You're already making an offering to me, because I'm greater than you ... and, therefore, wiser!" he says, winking at her in an attempt to get back to their previous conversation. "I'll tell you a story you're going to love."

"I'm all ears."

"First try the food." He hands her a pair of chopsticks. "Looks good, right?"

After a few clumsy attempts, she decides to grab one of the white squares.

"Mmmm, delicious, it has a texture like ... is it made of rice?"

"Impressive. Maybe you'll convince me that your famous sensory stimulation works; not many people would have gotten that right." He stares at her to get her attention. "If a Korean person were to ask you right now how many times you've eaten *ttokkuk,* what would you tell him?"

"Is that a trick question? You'll stop believing in stimulation if I mess up?"

"That's got nothing to do with it, woman; go on, take a risk."

"Once, I suppose ... or never, because I've only just tried it."

"Whoever you're talking to will wet themself laughing, or maybe they'd think you were making fun of them and be offended. They're asking you how old you are and you're telling them you're one year old ... or less!" Sebastian can't help but laugh. "I'm sorry, that's reminded me of the joke about the tourist who boasts about having climbed to the peak of the Himalayas ten or twelve times, and his friend, so as not to be beaten, tells him he knows he's been too, but he can't remember if it was once ... or less."

All this hilarity doesn't amuse Silvana.

"And can you tell me what age has got to do with all this?"

"That's what I wanted to tell you." He tries to make amends by putting on a serious expression. "The *ttokkuk* was only ever eaten on New Year's Eve, and the polite way to ask a person's age was to ask them how many times they'd eaten it. Someone told me earlier, when I went to get some, don't think I'm all that wise. What I do know is that Korean people would celebrate their birthdays all together on this night, so ... congratulations!" He raises his glass and holds it to her lips for her to drink.

Even though it was only a tiny sip, the burning sensation in her mouth will last for a good while. "I wouldn't know how to tell you what this potion is made of."

"Don't worry, you've got plenty of years ahead of you to work it out."

"Here we go again! You're obsessed, aren't you? Are you really that worried about the passage of time? Or is it because today is the day to talk about it, and now you'll bring up favorite old cliché, that one about how the older we get the more time flies ..."

"Cliché or not, there's truth in it."

"It must be that we're less bored every year." Just as she says those words, she is ambushed by the thought that's been obsessing her, that bodies and skin bore her. Perhaps everything will be different from now on, who knows, maybe she won't be able to converse so lightheartedly.

"You and your theories." He's also started eating from the same bowl, which is sitting between them. "It seems to me that when we were young we were impatient to achieve one thing or another, and it felt like we'd never get there ..."

"And have you already achieved everything you wanted to? I'm still impatient ..."

"You know what I mean, it's not the same. Let me put it this way: The days fly by because we're already familiar with every nook and cranny so we can move around without bothering about them too much, just like when you walk a well-known path, it feels shorter every time."

"You see, that's something we agree on." Being familiar with the nooks and crannies of the body helps her to tour around them in her mind automatically. "Routines prevent us from distinguishing one day from another, one body from another, therefore life passes us by without us realizing. It's important to flee from routines, from bodies, then."

Their bowl of *ttokkuk* has been empty for a while when the boy who's collecting them informs Silvana and Sebastian that they have to clear the area, which is next to an exit, as the dragon and lion dances will be starting soon. They will be scaring evil spirits away, and their retreat must be facilitated.

"I suppose this short time must have felt really long for you, but for me ..." He links arms with her and, before deciding which way they should go, kisses her cleavage, which is so well highlighted by the silk dress.

Taken by surprise, Silvana trembles, and, bit by bit, a smile dances across her lips, maybe seeing bodies bores her, but she's still sensitive to certain kinds of physical contact.

Walking slowly with their arms around each other, they head for an area that seems to be the source of shouts and general clamor. If they haven't seen any children up until now, it's because they were all concentrating on flying kites. The young adults, however, must have decided to celebrate the party outside of the ComU. The artistic beauty of the ceiling, a collection of shapes and colors outlined against a translucent surface and lit from above, contrasts with the confusion below. Kids of all ages, each gripping their string, are trying to bring down their neighbors' kites. It's the final battle. Silvana, surprised, realizes that some of them have stuck pieces of glass and metal to the kite to make their attacks more efficient and, indeed, when she looks down there are a fair number on the floor that have been taken down.

She feels a tapping on her shoulder and, when she looks up, she sees Justina, a star student from the stimulation course, who now works organizing kids' parties.

"I'm shocked," Silvana challenges her. "How come they're allowed to be so violent? It's against the rules."

"Not you as well... since this activity was proposed, that's all I've heard. If we're going to bring back a tradition, we have to reproduce it exactly as it was, right?"

"Maybe you could have chosen something different..."

"Look, things have changed, kids are different now than they were a few years ago." She's becoming so agitated by the topic that she's turning red. "Look at them, to me it seems healthier for them to attack each other than to ignore each other, which is what they tend to do these days."

"What do you mean?" In order to match Justina's gesticulations, Silvana lets go of Sebastian, who is paying more attention to the kites than the conversation.

"The other day, at one of those previous-century adoption parties, I couldn't get the kids to play together. They were all doing their own thing. I'd seen it before, but not this bad. They don't even look at each other anymore!"

"Tell me about it. I fight every day to make young people look beyond their bangs, but until now children weren't so damaged."

"Maybe not babies, but the ones I had the other day..."

"How old were they?"

"Of course, they were older." The girl looks relieved. "Thirteen."

"You mean to say they've unfrozen a thirteen-year-old boy?"

"A girl."

"Animals!" As she says it an idea pops into Silvana's head. "Listen, I'd like to meet her. How could I get in contact with her?"

"I didn't know you were interested in the unfrozen ... If I'd known ... the girl's mother asked me to recommend a psychologist and I put her in contact with Amalia, but if you want ..." Two little lads are calling over to her, brandishing their tangled strings.

"I'll talk to her myself, thanks. And if you hear about any other cases ..."

"I'll let you know, don't worry"—she practically shouts this as she walks away from Silvana to tend to the boys and calm them down.

Baltasar must have been waiting for the conversation to end to jump in, because, without her realizing, he's already between her and Sebastian with his arms around them just like before.

"I've done it! A present for you, Silvana: next year we'll have a Hebrew celebration. Since you're so interested in distinguished Jews these days, I guess you'll be very happy ..."

She feels rather out of place, not to mention Sebastian, but both of them allow themselves to be dragged along by this boundless torrent of energy and soon they're standing before a huge window watching the fireworks that welcome the new year. A year in which, surely this time, they will stop the boomerang.

12

Even though she's chosen to spend the big night at the gym, Lu has turned down the offer to compete in both the best proportions category and in the muscle mass/fat ratio category, not without a little regret, because last year she'd reached the final, and, this year, with the changes in the age categories, she feels she'd have a good chance of winning. But her daughter must come first. She wants to be there for her, introduce her to everyone, and see what everyone has to say on the matter, especially Fi.

As soon as they're inside, she sends ROBul to find a good spot for the four of them: "If possible, near the judges' booth," she orders. Taking care not to interrupt the girl's game with her robot—what a stroke of luck, that they get along so well—she searches the room for a familiar face. There are more people than in previous years. She notices a little blonde girl, very pretty, whom she's sure she recognizes from somewhere. She must be more or less the same age as Celia, maybe she's seen her at the school, no, now she remembers: she's the little princess she was so enamored with at the welcome party; in fact she's the only thing she enjoyed about the event.

Now she does interrupt the game to show her daughter who she's found and, while trying to point out the exact place to Celia, she realizes who she's with—she can't believe it—it's Fi! How has she managed it? She was so desperate, she might have asked someone to lend her their daughter, but who would have agreed on a day like today?

She's staring so much that her friend has spotted her and, judging by the gestures she's making with her arm, she's talking to her. And ROBul's not back yet! She leaves Celia to one side and talks directly to the girl's robot.

"You, ROBbie, finally you'll be of some use to me. Pick up what that woman is saying."

After turning his head left and right, up and down, in order to focus on the woman precisely, he plays back what she's saying:

"I've sent my ROB to find yours so they can find a spot for all of us. Have you heard from them?" Taking advantage of the pause, it turns and adds in its own synthetic voice: "ROBul has just sent me a signal with their location."

Celia has been watching this show in awe—she can't work out how he could have picked out what the woman was saying from among so much noise and other conversations—but she saves the question for later, when they're left alone. Now she has to be there for Lu, who, after telling the robot to show them the way, has taken her hand and won't stop tidying her hair and dress, and stroking her affectionately the whole time, although that doesn't prevent her from waving to people and introducing Celia to everyone who comes near them. She also asks about the blonde girl:

"Have you seen her since the party?" she asks to hide her anxiety.

Celia looks utterly confused.

"Yes, every day at school, she's my best friend."

"And how come you haven't told me about it?"

Lu squeezes her hand so tightly it hurts, making her even more confused.

"I'm always telling you about Xis. Don't you remember?"

"Ah, yes, yes, sorry." She tries to diffuse the tension. "I just didn't know it was the same Xis who came to our house. And who's her mom?"

"I don't know, I've never met her."

The spot they've ended up in couldn't be better, next to the arena and overlooking the whole auditorium. Moments after they arrive, Fi appears and starts praising ROBul. All the pride Lu would have felt on any other occasion is turned into anxiety and, in an attempt to reinstate her daughter as the center of attention she deserves to be, Lu takes hold of her daughter's shoulders and gently pushes her toward Fi until she's standing right in front of her.

"Here she is, my daughter." It almost sounds like a challenge.

"Yes, I've seen her. Hello, darling." Hardly having taken a look at her, she turns toward the other girl and adds: "This is Xis; I thought you could entertain each other while your mother and I compete. I suppose you're not yet… familiar with playing with ROBs." She has to shout the end of the sentence in order to be heard over Lu's protestations, who tells her to stop talking about things she has no idea about, by which she means the competition, and also Xis' shouting, as she does want to participate and doesn't intend to play the role of ROB for anyone.

The rise in volume makes the respective ROBs draw closer in case they have to intervene.

"We know each other from school," Celia points out, surprised by the introduction, and, most of all, by the aggressive replies. "Don't worry about me, I always have a good time with ROBbie."

Xis ends up getting her own way and manages to get Fi to hand over her registration, despite the committee's reservations about the change of category. Accompanied by her ROBix, she happily heads off to the changing room to get ready, carefully watched over by Celia, who's waiting for some kind of farewell gesture she can respond to.

"It's better to let her be happy. She gave her mother enough trouble this morning." Resigned to not being able to compete, Fi lies down in a hammock next to Lu's and selects the same relaxation and well-being sequence her friend has chosen.

"Do you know the mother?" Now that the little girl's not there, and while they're waiting for the show to start, it seems like a good moment to clear up a few things.

"Sus Cal'Vin, she works at CraftER; I did tell you their all-powerful president exchanges riddles with Hug, right?"

"Yes. And how come… ?"

"Sus had to accompany an important magnate to the convention and, obviously, she couldn't take her daughter. Apparently she's never let her go and this year she'd promised. The kid throws the most spectacular tantrums…"

That's why she's so strange, Celia thinks to herself, having refrained from asking ROBbie about that incredible thing he did before so that she can listen to the conversation.

"And is Hug there too?"

"As if! He doesn't want anything to do with Dr. Craft's business. He says if he's anywhere near as dangerous there as he is in their riddle duels... He's gone to the dueling club, for a change, and then he'll come and pick us up." All of a sudden she remembers the last time they met up with the Doctor. "Have you already found out which dog you'll look like when you get old?"

"What on earth are you talking about now?"

"He gave us a real shock the other day. Suddenly, in the mirror, I saw myself transformed into a poodle. I swear: it had a face just like mine and it moved exactly when I did. Next to me, Hug had turned into a boxer. Yeah, sure, laugh it up, I'd like to have seen you, with hair all over you, a wrinkled up face and no eyelashes. It was horrible. Since he's better at technology than anyone else, he can really trick you. He has such a sick sense of humor, it's practically sadistic."

Celia doesn't see any of it as being sadistic, maybe it is for the dogs, now that she thinks about it, she's only ever seen them in virtual images and mechanical replicas. What must they have done with the ones made of flesh and bone?

"Speaking of wrinkles"—Lu grabs hold of the first thing that comes into her head to get away from so much salaciousness—"have you heard there's a new treatment to get rid of them?"

"What does it matter to you? You haven't even got any!"

"But you have. Since you did that course at the ComU..."

"Don't remind me! First of all, all that emoting gives you wrinkles... and then someone starts offering to eliminate them. It must be a conspiracy."

"Don't speak nonsense; they're on opposite sides."

"You *are* naive. Is this new treatment you're talking about IFC?"

"I think that's what they call it, yeah."

"So, I think you should know that IFC stands for inverse facial conditioning. It's not cosmetic, or surgery, or any kind of gadget... it's pure psychological therapy! You see, they've gone over to the other side. I'll stick with injections thank you very much."

Fed up with not understanding much and bored by what she does understand, Celia asks if she can go for a walk with ROBbie. That's when Fi makes the most of the situation to go on the counterattack and bombard Lu with questions. Why had she accepted such a grown-up girl, what will she do if she gets pregnant, or maybe they've sterilized her? What a drag, having those period things, does she already get them?

Lu doesn't even try to answer, she simply orders ROBul to close her auditory eyelids and add some more tranquilizer to her sequence. She knew her friend's desire to adopt a child was held back by her cowardice, but she couldn't have imagined it had gotten so excessive.

Once they've put a few paces between them and the two women, Celia can ask her question:

"How did you pick out what Fi was saying with so much noise around?"

"*Information*: I have a source separation program installed."

"Sauce?" She doesn't understand what condiments have to do with all this.

"*Definition:* sound emitters."

"Oh, I get it, you mean the source of the voices. And does the program allow you to separate each one of the voices that are all mixed up when you hear them?"

"*Correct*."

"But how does it work?"

"*Access denied:* I can't provide that information."

"You can't or you don't know?"

"*Not understood:* It's the same thing."

"Of course there's a difference, ROBbie, I told you about that the other day."

"*Recovering the other day:* So, just like then, my answer is that I have no other master apart from you."

"But maybe it's you who doesn't want to tell me."

"*Impossible:* If you wanted to know, and I knew the answer, I would tell you."

Celia stops for a moment, touched by the words, and looks for his eyes: no friend had ever sworn their loyalty so convincingly, but two black holes bring her back down to earth. Though not entirely. As they start moving again, she watches the robot out of the corner of her eye and it pleases her to see his dignified posture, gently swinging his strong, shiny arms. It feels good to walk along beside him, she feels protected, she can trust him. And what does it matter that he doesn't have eyes, people don't look at each other anymore anyway.

"*Suggestion*: Why don't you ask me how many voices I can separate, or from how far away? Or, simply, use me to spy on someone you're interested in? You'd enjoy it more."

"The people I'd like to spy on are really far away. You wouldn't be able to hear them."

"*Information*: I have a range of 1,500 feet. Shall we try it?"

A sad smile plays across Celia's face.

"Let it go, ROBbie."

"*Recommendation*: Don't get angry. I told you I can do it, but I don't know how to explain it to you. You know how to stand on one leg: I saw you do it the other day, and I bet you couldn't tell me what sequence of muscles you use to do it."

"Good example, yes indeed. Do you know how to balance?"

"*Negative:* Not on one leg, that's why I was surprised."

"Well, I don't know which muscles are used, but I can show you how to do it."

"*Attention*: I can see what you're getting at, but it's different. You don't have the necessary organs for source separation."

"How do you know, if you don't know the mechanisms you use to make it work?"

"*Analogy*: Because you can't use your eyes as a telescope either, and the mechanism is the same: isolating a small area and magnifying it, whether it's sound or vision."

"So you do know something about it. I thought you had like a sound magnet that only attracted the voice you wanted, but now you're telling me you capture all the background noise and then you isolate the voice you want. Is that right?"

"*Repetition*: I don't know. And I don't understand why you want to know how it's done… you've got me to do it for you."

"ROBbie, I'm happy to have you, believe me. It's just that, since I was little, my father taught me that I had to be able to look after myself, and not depend on anything or anyone. And in the girl scouts, our war cry was 'AUR,' autonomous and responsible. Can you understand that? Although, maybe I am taking it a bit too far, I'll never be a telescope."

She waits and watches to see if there's any indication that he's amused. She gives up. She doesn't like to think that he'll never have a sense of humor, so she decides it's just because he hasn't gotten the hang of that type of joke yet. She hasn't managed to make Lu, or Xis, or anyone else laugh so far.

"*Offer:* I've got a zoom that can magnify almost as much as a telescope. Who do you want me to focus it on?"

"It's not right to invade other people's privacy."

"*Defense*: It's what people come to these parties to do; otherwise, everyone would stay at home."

"You mean they come to eavesdrop and be spied on?"

"*Correct.*"

"So the people down here are just as much exhibitionists as the people on the stage? Lu and Fi as well? It looked like they wanted to talk about things."

"*Information*: People who want to protect themselves from eavesdroppers use encrypted electronic communications."

"And me and you right now?"

"*Concession*: It's possible someone is listening to us."

"Listening to us? Who?" She really didn't expect this.

"*Access denied*: There's no way to know. It could be ROBul, if Lu's asked him to."

A sudden feeling of insecurity washes over her.

"And you, has she ever asked you to tell her about our conversations?"

"*Impossible*: She knows I only obey you."

Despite its forcefulness, the answer doesn't make her feel much better this time. And as soon as they're back next to Lu, she can't wait to bring it up:

"Is it okay to listen in on other people's conversations?" Faced with Lu's confused expression, she explains herself: "I don't mean like before, when Fi was talking to you, but random conversations between people who aren't talking to you."

"Ah, you like to eavesdrop on what they're saying? Of course, darling, if it makes you happy, it's fine."

She almost tells her she's misunderstood, that she doesn't enjoy listening in on strangers, actually it makes her feel uncomfortable because she feels like she's up to no good; that's what she meant, she wanted to know if she felt that way too or if she did it... But it doesn't matter, the reaction is clear enough and it gives her an answer to both questions at once... and too many others. If what makes her happy is good, what makes her sad is bad? Like a desert that opens up in the middle of a larger one, or a tiny island emerging from the closed waters within a larger island, Celia feels that a new solitude, insoluble and very different, has superimposed itself on top of the one she is already experiencing in this new century.

III
THE CREATIVITY PROSTHESIS

13

Since Leo has been working on the E-Creative project, under the Doctor's supervision, he's been assigned an individual cubicle in the most exclusive wing of CraftER. He's heard a lot about this part of the building. His ex-coworkers tried to hide their envy by citing the rumor that if it was difficult to get in, it was even more difficult to get out. As if someone, after being promoted, would want to go back to their old position, he argued back when they started clutching at straws, citing all kinds of pretexts in order to ignore the evidence. If people were spending whole weeks without coming out it must be that the level of comfort and the rewards inside were incomparable with those outside. And if, once finished, they left the company, it would be to get even better jobs. He refused to listen to fantasies about brains pushed so hard they went mad, thrown out once they were no more than human waste, and he was even less prepared to believe them.

Now he's been able to confirm his suspicions. It's been days since he went home, not even to sleep. Why should he leave if he's got everything he needs here and it makes him happy? Everything apart from Bet, of course, and maybe chess matches. He himself cannot quite understand his addiction to the physical presence of his opponent. Long distance, he's lost matches against adversaries that he'd always beaten and, worse still, he hadn't enjoyed the victories, as if it were a very different game than the one he usually got so excited about.

First his old colleagues, then Bet, and now chess ... Why can't he concentrate today? At this rate he won't even finish the program he left practically completed last night. He must be unsettled by the prospect of leaving the cubicle so soon. Like it or not, knowing that all the information about the prosthesis will be erased as soon as he crosses the threshold makes him a bit nervous. Dr. Craft had assured him that, apart from that memory lapse, he wouldn't notice anything else. In fact, he gave him a practical demonstration when, terrified by that clause of the contract, he was about to change his mind. The timeout button, as he called it, was just a simplified copy of the mechanism incorporated into the cubicles, but it really did the job. He can clearly see in his mind's eye the dueling table that the Doctor had proudly shown him: its wooden shell with swords set into it and, in the middle, the spectacular touch screen. He can remember perfectly that a riddle popped up on it and he read it, and, what's more, he found it absorbing; but as soon as he pressed the button, it was erased not only from the screen, but also from his memory. He couldn't have explained how it worked even as a matter of life and death.

The waves of encryption the device added to the brain were innocuous, he'd checked it out. There weren't any side effects either at the time or after, so in that sense he isn't worried. What annoys him is being at the mercy of a mechanism that he doesn't understand. Although the Doctor explained the basic idea to him, hinting that he could work the details out for himself, he hasn't been able to get his head around it and has ended up convincing himself he's been conned. Because ... who would reveal the secret that gave them an advantage? Not the president of CraftER or most people he knew, starting with Bet. How many times had she warned him to take his inventions off the net before the company ordered him to do so? He must be the only idiot who gets a kick out of divulging his inventions. As impractical an addiction as needing the physical presence of an opponent when playing chess. If he at least had the table available to him, he could run tests and try to find out how each component worked.

He still doesn't understand why they call it "timeout," since it's just a memory-less time that passes like any other. Maybe it's a timeout for the duel, as it's postponed, but it's definitely not for the players. That's something that hasn't been invented yet and would be something he'd like: That the world carried on its normal course with him able to pause and then un-pause later on when the great technological fantasies have been brought to fruition, with him able to make good use of them. Or even the opposite, stop the world so he can spy on what's being done in other laboratories and research centers, without anyone stopping him, without even realizing, since he'd be going at infinite speed compared to the poor people stuck on pause.

At first he thought the Doctor's device would be useful for his wireless transmutation. At the same time that one person is injected with the electroencephalographic records of another, they'd be emptied of their own records or, if not, have them shut down to avoid interferences. But straightaway he realized that it didn't work like that, the timeout device didn't take memories away, it was more like adding them: it emitted waves of encryption that attached themselves to the waves generated by the individual when they read the riddle or performed any other activity. To erase the record, you just had to deactivate the transmitter. The signals generated in the brain, even though they're the same ones as before, become mere noise once they're stripped of the encryption base. The memory is there ... it will be there when he leaves the cubicle a couple of hours from now to go pick up Bet, but he won't have the key to decode it.

Lost in thought, he hasn't even realized that ROBco has taken control of the wrap-around screen that, when stretched to its maximum, covers the walls of the space, and when it asks permission to deactivate the holographic partition walls, he almost jumps out of his seat. It has the results of the comparison it was assigned this morning; not like him, who hasn't been able to stop getting side-tracked by his imminent outing. How little control, he's making a fool of himself, and this moron, with all its neurolearning and pedigree, hasn't even learned to stop him when he's wasting time like an idiot. He looks up at the ever-watching electronic eyes of the cameras and thinks how lucky he is that they can't read his mind. Then, the idea that his dreamed transmutation could make that kind of mental surveillance possible briefly crosses his mind, but he doesn't make an effort to retain the thought.

He has before him, in parallel, representations of the structure of a cutting-edge hypothesis generator and the imagination algebra he's been working on. He's deliberately developed it without knowing anything about the generator so that, a posteriori, he can compare them and extract the best of both.

As expected, the generator follows an evolutionary scheme, consistently applying random mutations to the same idea and selecting the best variant. As if blind Darwinian evolution could be a font of creativity! He would never rely on fate doing his work for him, not when there are powerful automatic inference tools to guide the choice of possibilities. Thanks to these tools, his algebra mechanizes the creative process, avoiding all randomness and subjective evaluation. He's pleased with it. But he has to admit that it has a limited, determinist scope. It's possible that by adding some random but controlled sources from the generator he could increase its power.

With the help of the panoramic chart he's able to quickly get a global picture, but meticulously analyzing which mechanisms are worth importing to his algebra would take hours. Plus he isn't in the mood right now, even if ROBco is waiting expectantly next to him, ready to act as his assistant.

"It's not worth us starting, if I have to leave in a moment." As ROBco is still staring at him insistently, he admonishes it, "I told you: I can't turn off and back on again and pick up where I left off, like you do, see if you can finally build that into your model."

"*Confirmation*: It was incorporated seventeen days, four hours, thirteen…"

"Stop, stop, stop… I've also told you several times that it's not necessary to be so precise. And if you're aware of my limitations, I don't understand why you insist on starting work."

"*Playback:* You said "I have to leave in a moment." Question: why?"

"I get it: I forgot to inform you that I'm going out with Bet today. Okay, you don't have to correct anything, my mistake."

"*Request*: What do I have to prepare?"

"We'll take the aero'car. The Doctor assured me there'd always be one available."

"*Acceptance:* I will check it is ready. *Consultation:* any other tasks?"

"Yes, it would be advisable to get ahead with the work. The next step will be to fill this formalism with content"—he waves at the diagrams covering the walls. "Collect as many encephalic records as you can from people performing creative tasks: artists, designers, inventors… and also mathematicians and physicists demonstrating theorems, to see if I'm right in thinking there are similarities. Use the same amount of geniuses as anonymous people, above all we need a wide variety of data in order to process it and extract correlations. It would be best if the records had introspective annotations attached."

"*Workload accepted.* Any additional criteria for candidate selection?"

"It's crucial to avoid statistical normality. Perhaps it would be wise to include a mentally ill person within such a rich universe, but mainly people without prejudices, not confined by any rules, people who hate routine and are averse to repetition. Keep in mind that a creative spark not only depends on individual factors, but is also fostered by shock: between cultures, environments, traditions. Look for people who, for whatever reason, have been transplanted to a different environment. They say that a genius is a person who moves forward, that looks and lives beyond their time."

"*Consultation:* Do I have to search on the web or do you want me to travel to some data warehouses?"

"Start with the web. If later on I need to send you..." Suddenly, a thought leaps into his mind. "Can you save everything on the screens now to the absolute memory?"

"*Denied:* It is prohibited, I thought you knew."

"Can you save it in a perennial storage device, then, or even in a compatible peripheral?"

"*Denied*: Information belonging to CraftER cannot be copied to foreign storage devices. Why do you want it?"

Instinctively, Leo looks up to the surveillance cameras, without realizing that the movement, along with ROBco's words, sets off the possible incident alarm.

Just ten minutes later, a virtual image of the Doctor is plonked down before him.

"Hello, Leo. Is everything ready for the demo of the preliminary version of the prosthesis tomorrow?"

"Good afternoon, Dr. Craft." He tries to hide his surprise at a visit that was not drawn up in the control schedule. "Not the demo, but I've got the mechanism diagram finished. Would you like to see it?"

"Don't bother showing me pipe dreams, I want results. Development by layers and a demo for each layer. I made that clear enough, didn't I?"

"Yes sir, but I can't add content without having the overall scheme ready."

"And I can't wait for you to finish the project before I decide if you're worth anything to me or not."

"But incremental design is slower, and often produces a patchy final product."

"Cut the spiel. The others have already sent me prototypes; limited ones, of course, but still of some use to me."

"Others? You mean there are other people working on this project?"

"What did you think? That you're the chosen one, the golden boy?"

He'll always be naive. If Bet were here, she'd tear him apart. He'll never tell her about this.

"Fire me, if you want." He builds up some courage. "I know if you take a look at the overall design you won't demand useless demos that could slow down the production of a decent prosthesis."

"You're so proud of your design that you intend to sneak it out of CraftER, I see."

Leo feels blood rushing to his face, and lowers his eyes, away from the screen.

"It was a passing thought, you must have seen it wasn't premeditated."

"That's what's saved your skin. I don't care if you break the rules, creative minds are like that, what I cannot tolerate is you wasting time on this bullshit. Get it into your head that it's impossible: the timeout device is one of the projects I directed personally."

On the screen, the arrow formed by the Doctor's eyebrows appears blurred and doesn't seem as threatening as in person. Just like his words, which seem to come out agreeably: he even recognized that Leo has a creative mind, the highest compliment you could imagine coming from Dr. Craft's mouth. People really are unpredictable; he's caught out, and then, as part of his telling off, he's pummeled with praise. He's been left in such a vulnerable state that, without realizing, he's already agreed to have a demo ready for the next control meeting, ten days from now.

The sensible thing would be to get down to work straightaway and cancel his outing, but Bet would never forgive him. She's already enough of a pain in the neck with her insistence that he's deliberately keeping the project a secret from her. It'll be difficult to make her understand that, outside of CraftER, it's not that he doesn't want to explain it to her, it's that he can't remember what he does when he's in there. With some exceptions, of course, he hopes to at least be able to keep her happy by telling her about the everyday details; the Doctor had assured him that the amnesia would be limited to the project.

He's never been so aware of himself as when he crosses the threshold of his cubicle. He does it very slowly, apprehensively, blind and deaf to everything around him, because his eyes are closed and he's listening to his insides. He watches over every little feeling, every tiny beating of his restless pulse. He's afraid his head will explode, that he will be unable to hold his train of thought, that he won't recognize himself…he's not entirely sure what he should be afraid of, where he should focus his suspicions. And then he's already on the other side, without noticing anything strange, and now he looks down at his legs, his feet, he feels his arms, searching for external signs of a change he was unable to recognize under his skin.

Only when he searches his mind for the details of what he has been developing does he find an impenetrable empty space and is able to confirm that the Doctor hasn't tricked him. Oh the irony, the emptiness that worried him so much is now a relief and, feeling more relaxed, he heads for the platform where ROBco is waiting with the aero'car.

It's a spectacular two-seater, worthy of CraftER, stylish on the outside and spacious on the inside, with cavities to fit their respective ROBs; Bet will be delighted.

He was right, when he picks her up, Leo has the privilege of witnessing the biggest smile he's ever seen on her face, and he momentarily questions his intention to cut their outing short in order to return to his demo. Tightly squeezing both her hands, he helps her into the front seat so that, face to face, they can plan their time together. It's what they've always done, the secret to their successful relationship: deciding in that moment and in light of their current mood, without being held to a preestablished script dictating what they should do.

"Are you feeling desire today?" They always start with this, as it affects everything else.

"This morning I was at 6.7, but I think it's dropped. You?"

"Nothing remarkable. Let's forget about going to the health club, agreed?"

"Yes, yes, I'd prefer to make the most of this wonderful vehicle. What is it capable of?"

"No idea, ROBco flies it."

"*Information*: I have been in charge of it for ninety-five minutes, I have only mastered the basic controls."

"Ohh..." Bet doesn't hide her disappointment, rather she exaggerates it.

"What if we go for a short flight today and save all the experienced piloting tricks for next time? I promise that ROBco will dedicate himself to it one hundred percent."

"But I feel like doing it today. And isn't your ROB tied up with the top secret project?"

"Don't be so acerbic."

"I'm only telling you what I think and what I feel, that's what we agreed, right?"

"Okay, so I will too"—she's handed it to him on a silver platter—"the president has just assigned me a demo and I'm in a hurry to get back."

"Couldn't you have told me that before?"

"You would have erased me from your contact list indefinitely."

"Well I will if you don't tell me what your project is about."

"I don't know, when I left they made me pass through a device that erases my memory."

"Don't mess with me, Leonix, this is our big joint project." Rather than being angry, her expression is serious, as if she doubted the sanity of her partner.

"Joint?" Stunned by an expansive wave of doubt, he mechanically repeats the final word. The panic provoked by the effects of timeout becomes confused with the fact that he has no idea which project she's talking to him about.

"Don't look at me like that! You're scaring me. Our project, don't you remember? Or have they erased that too? It's about selective memory loss." As she sees her words are bringing him back to normal she continues, "You baptized it the 'happiness app,' is that ringing any bells?" She moves closer to stare at him, somewhere between amused and alarmed. "By adjusting the filter between what we remember and what we forget, we would be able to create a fortunate past for ourselves. You even talked about taking it further and introducing fictitious, very pleasing memories..."

"Yes, of course I remember." He was so frightened by the ghost of the forgotten that it hadn't even crossed his mind that she might have been referring to that other project. What a weight off his shoulders: he knows everything they've said about the app by heart; and more, he also has the details he's been polishing on his own. "But that's got nothing to do with CraftER's protection system."

"Hasn't it? Didn't you say they've erased certain things from your memory?"

"If you put it like that, then yes, but the principle is totally different. They can erase it because they've saved it in a special way, with an encryption base. Our app is much more ambitious: it has to be able to erase things without any control over the recording process; past events, for example."

"You mean you haven't given them our designs?"

"Of course not, sweetheart, it's a different project, it was developed before they hired me."

"How can you be sure that's not the part of your memory they erased?"

"Because, when I started, they demonstrated the security mechanism to me."

"Maybe they've gone one step further than us and they know how to introduce fictitious memories."

"Why do you always have to be so suspicious? What's it going to be: that they've gone a step further or that they've copied us?"

"You see? I worry about you and you get angry. Such an intrusive protection system must be illegal. There's one at MascotER too, but it doesn't put the rights of employees at risk."

"How would you know how they work on exclusive projects!"

"That's it, the great genius has to speak up."

"Well yes, you know what, I prefer to move forward than to be stuck in a rut. All this reticence stops you achieving anything. Look"—he glances outside—"we've ended up staying here, we haven't even had a little run round the block."

"It wasn't our day today, that's all." She gets up, determined to get out of the aero'car. "You were in a hurry, there's nothing more to talk about. Just tell me beforehand next time, and we'll just cancel."

Leo, ever the gentleman, again takes her by the hands to help her out, but whereas before her smile made him question his will, now nothing can take his mind off throwing himself wholeheartedly into the demo.

14

It's been a long time since Silvana last left the ComU, a couple of years or more, so it's no surprise that everything's changed almost beyond recognition. Luckily, Sebastian has registered the visit as a service of the Center so she's been assigned a driver, otherwise she would almost certainly have gotten lost. In the past you could count on one hand the number of buildings in which you could fly around in an aero'car—the general hospital, multistory parking lots and not much else—but now flying outdoors is the exception to the rule. Maybe it's for the best, the brightness bothers her, she's not used to it anymore. Of course, with a ceiling overhead the feeling of entrapment is heightened, with so many vehicles and landing platforms that it's claustrophobic.

It was right under her nose all the time but she hadn't realized that the buildings grew in layers, adding on residents until they took on the form of these half-mile blocks. If they're shocking for her, imagine how shocked the poor little girl must be. She's confident that she'll meet her today, despite not having been able to set up anything concrete with the mother. Most of all she must remember that she's been hired as a therapist; her personal interest in the potential archaic feelings of the girl will have to stay hidden, she can't show it at any point during the interview.

As she'd guessed, the home is located in a well-to-do neighborhood, and the entrance the driver points out to her is one of the nicest ones around: on a high level with a private platform. Once Silvana has connected with the owner's ROB, she receives free license to dock there and, if she wants, she can leave the aero'car parked there until it's time to leave.

She's not in the habit of visiting people at home and feels a little unsure of herself as she walks up to the house, not knowing how to greet the smartly dressed client waiting at the door. But Lu, before Silvana has reached her, is already going back into the house and beckoning for her to follow. Most of the walls are yellow and the furniture is so dazzling that it's like being back outside again. Worse still, the suspended armchairs they sit down in are the same color, and they're so hard to look at she has to find some kind of neutral space where she can rest her eyes, which she finds between her feet.

The obvious nervousness Lu demonstrates while she explains her daughter's eccentricities puts her at ease: she doesn't need to worry about the impression she might be making, she's clearly not making one. It's hard enough for this woman to string a sentence together, let alone pay attention to someone else at the same time. She begins to think they might as well have exchanged information over the net, and that it'll only have been worth coming if in the end she gets to meet Celia.

As soon as she gets a chance to ask, she finds out that the school is nearby, and that the girl will be coming home in less than half an hour. She can wait for her, if she wants, or she can connect with the monitoring circuit reserved for parents and observe what she's doing right now. Silvana hesitates for an instant. She should decline the offer without a second thought: she's denounced technology that violates privacy so many times, for how it undermines the privacy of the weakest among us; if someone from the ComU were to catch her spying, she would die of shame. But the opportunity is too tempting to refuse so she justifies her decision by telling herself that observing a spontaneous situation will help her diagnose the problem and, at the end of the day, it'll be for Celia's own good.

A few minutes later ROBul informs them that, as the girl isn't showing up on any of the static cameras, they've sent a SEEKer to look for her. Lu can't believe it, after she had insisted that Celia try to be like the others and slip under the radar. Her teacher had warned Lu that if she didn't adapt soon she would have to change schools. It occurs to Silvana that if you want to slip under the radar, what better way than hiding from the cameras, but she doesn't say anything, convinced that Lu wouldn't take it well, and maybe wouldn't even understand. Gratuitous humor wasn't exactly widespread in the pro-techno community.

Finally the SEEKer sends them some images of Celia. She's in a large space that Lu identifies as the socialization classroom. She has a wide-awake expression and light brown, very straight hair tied back in a ponytail, and looks just like she expected. The only difference is that she hadn't imagined her with freckles, you don't see faces like that anymore; strangely it gives her a touch of mischief, as if the freckles made her brighter, smarter. Next to her, also sitting on the floor, a girl with short, curly blonde hair is hugging her knees, and appears to be crying.

"Oh no, we're going to end up fighting again," Lu exclaims, for once forgetting about her airs and graces. "Who knows what she'll have said to that little angel."

"Can we hear what they're saying?"

"No, no. It would violate the rights of the child, you should know that," she says, giving her a suspicious look.

"Sorry, but I don't understand: the images are public but the sounds are private?"

"Come on! Who said anything about it being public?" She seems to be outraged, it must be a hot-button issue. "Parents have a right to check on the physical integrity of their children at any time, that's all. If someone breaks that rule, with lip-reading programs or any other tricks, their connection to the circuit is cut off forever." She pauses to catch her breath. "So tell me... what experience do you have as a therapist with unfrozen children?"

"Hardly any have been unfrozen at this age, I guess you know that. The experienced therapists have worked with babies and can't get a handle on this case. They've assigned it to me precisely because I'm a specialist in the emotions of the past"—she's already said too much—"and I'll be able to do more than a normal therapist. I sent you my credentials, did you get them?"

"Yes, but I assumed you'd dealt with cases like my daughter's."

"Look, the girls are standing up." Silvana hasn't taken her eyes off the monitor the whole time. "What lovely hair."

Lu's face tightens:

“She just had to let it down, didn’t she ... No, in the end they’ll make me cut it.”

It’s almost like Celia has heard her, just then she gathers her hair and fastens it back with a turban.

“What’s the problem?”

“Do you think she’s avoiding standing out behaving like that? I’m starting to see the experience you’ve had with girls this age.”

“Celia must be different in a lot of ways ... I didn’t know hair was so important,” she says with contempt.

“Think about it, it’s the first thing you noticed.”

Like a boomerang, her assault has come right back at her with total innocence. Instinctively, she changes the subject:

“What are those figures they keep bumping into, they look like mannequins.”

“They’re for practicing socialization.” She stops for a moment as if she can’t be bothered to explain it. “They stage a situation and the kids have to practice until they learn to behave properly automatically. It’s one of the most innovative activities in the school, they call it social-conduct training; they recommended it for Celia and it’s been good for her, in just a few days she’s caught up with the rest.”

It occurs to Silvana that it’s like learning to drive, only that instead of controlling a machine that navigates among other machines, it’s navigation among people that is automated, but again she holds back the comment. Who knows, maybe this practice isn’t so stupid, considering how bad things have gotten. Nothing could be more pathetic than the statues she’s encountered recently in her classes; maybe mechanizing certain things would unblock them.

The monitor is switched off and ROBul communicates that Celia has left school and, taking into account the state of the traffic, it estimates that she will be home in seventeen minutes.

“Perfect, I can wait until she gets here,” Silvana jumps in, anticipating a possible intervention on Lu’s part.

“We agreed the therapy would start next week. I haven’t said anything to the little girl yet.”

“Don’t worry, we’ll start the sessions like we said, today will just be an introduction. And do me a favor, don’t say anything about therapy. She must have been through enough already with her illness without people calling her ill again now."

"How should I introduce you then?"

"Leave it to me." She's taken control of the situation and, for the next few minutes, she hopes to push the conversation toward what she's interested in.

This way she finds out that the teacher has labeled Celia a rebel because, ignoring his advice, she insists on competing with machines. She wastes time playing at being a calculator and a spellchecker instead of making an effort to learn the content she's been assigned. Among her classmates, of course, she's become known for her skills of deduction, which has made her a leader. But that doesn't help her academic attainment, which is very poor.

It was the EDUsys that started sending out alarm signals because she didn't use it as she was supposed to. Apparently she hasn't taken to the net's search mechanisms and, faced with a question, she stops and thinks about it, trying to make up an answer, instead of trusting what other people have thought before. "Imagine if we all had to start from scratch!" the teacher exclaimed, annoyed. He himself doesn't have most of the knowledge they are working on, he told her with pride, that's what EDUsys is for, lesson planning and doing the tedious work of evaluation, writing reports and keeping a record of each student. It also suggests to the teacher where he should invest his time in order to effectively facilitate interaction between the group and the system; but it's him who makes decisions, and the delicate subjects, like negotiating with other schools to hold joint sessions on advanced topics, are managed by him personally. It's a cutting-edge school, capable of reprogramming activities in real time according to how the day is going; she won't find a better one, and she doesn't want to look for one either.

Caught up in the academic issues, Lu won't let herself be pushed into the emotional and familial ambit where Silvana wants to take her. On the contrary, she makes it very clear that Silvana's being hired to help her daughter adapt to school. Nothing more. Just making her a little more docile will solve most of her problems. She has no doubt there is a great range of techniques available to mold character, and hopes that Silvana will be able to put them into practice. Indeed, she imagines Silvana will have to work hand in hand with ROBbie, because everyone knows that if the child turns out to be a rule-breaker, the robot must learn to restrain them, to counteract their impulses, put them on the right track ... Robots are customizable for a reason, they have to complement their PROPs to make a good team.

That's the last thing Silvana expected to hear, that she'll have to train a robot! It's really hard to bite her tongue and not let her different point of view spill out. She's afraid that, if she contradicts the woman, she'll be fired before she's started, and she would at least like to meet Celia. What an ordeal she must be going through with this mother, the poor thing. Or maybe not, in which case, she would have to give up the job herself. Although it's interesting for her investigations, there's always a limit.

She glances at her watch impatiently and, as if responding to the signal, ROBul comes in to announce the girl's arrival. They stand up at the same time, but only Lu goes to meet her daughter, having gestured to Silvana to wait a moment.

Perhaps because she's been staring at the opening for some time, partly expectantly and partly because she's seeking a refuge from the offensive yellow walls, or maybe it's because the dark silhouette that suddenly appears there is nice to look at, Celia's shy entrance, full of curiosity, makes a great impression on Silvana's retina and beyond. Without the turban and with her shining hair tied back in a braid, she looks taller and more slender than she'd imagined from the pictures. What's more, since she's used to seeing her students' rigid and inexpressive bodies, she finds the flexibility with which she moves fascinating. Behind her, Lu initiates the introduction:

"Silvana was interested to meet you, she's ..."

"I'm an emotions masseuse," she says, standing up. "It sounds a bit weird right? But we call it that."

Celia's black pupils seem to X-ray her, stopping her in her tracks: she mustn't make a wrong move.

"She'll help you with your studies," Lu intervenes, uncomfortable with the silence and making it clear that she does not intend to sit down.

"Yes, of course, but you'll have to help me out as well."

A note of naivety cracks Celia's serious stance, and the little girl she really is peeks through:

"Pardon?"

"You see, I'm interested in people who lived a long time ago and maybe you can help me to understand them better," she admits, ignoring what Lu may or may not think about it.

"Didn't you say you were a masseuse?"

"Yes, but that has a different meaning now…"—she searches for the right words—"than before you traveled through time, I mean."

She's obviously said the right thing, because, abandoning all her reticence, the girl comes closer and asks, curiously:

"You know how to travel through time?"

Lu's impatience has reached its limit.

"You can talk about that during your sessions. Silvana was waiting to see you, but she has to go."

If she wants to come back she would do well to follow this instruction, and it's come at an opportune moment, because she has no idea how to answer. As she moves toward the door, she notices that Celia wants to follow, but her mother stops her, and the girl resigns herself to a goodbye wave, which she hurriedly returns.

Against all odds, Lu doesn't let her leave accompanied only by ROBul; instead she joins them and, before Silvana gets onto the platform, she holds her back for a second: she wants to ask her something. Silvana didn't see this coming, before they had time to kill and now there's something urgent to talk about.

"I just got a communication from the clinic saying we've been invited to a get-together…"—she looks at her ROB's embedded screen uneasily—"an inter-century adoption get-together. Do you know what it's for?"

"Well, no… but I have some idea," she rectifies quickly. "It's good to meet other children who are in the same situation. For them and for their mothers. They tend to have similar problems, and seeing how other people have resolved them can help."

"So you would advise us to go." She looks concerned. "Could you come with us?"

"Of course, I'd love to." It would be an opportunity to make contact with other unfrozen teenagers, she thinks, before immediately withdrawing the thought, as if it were a betrayal of Celia.

After reserving Sunday two weeks from now in her calendar, confirming that they have four sessions before then, and finally saying goodbye to Lu, Silvana allows herself to think over what has just happened. What kind of commitment has she made with this child that could possibly stop her from meeting other kids? It makes no sense at all, but she felt it. She must have gotten so involved with obsolete sentiments that she's entered into their orbit herself. But her reading isn't the cause, it's more likely the black pupils that had captivated her, clearly communicating that all their fragile hopes rested with her. She feels she can't fail them and, at the same time, she feels she's in unknown territory where her own baggage won't be of much help. She's the newcomer here.

15

7:34 p.m. – I observe that the Doctor is still absorbed in the CraftER monitor. Forty-eight minutes have passed without him giving me any orders, not even to curse at me. I speculate that he may have fallen asleep. I move closer, making sure not to make too much noise and I confirm that he is still consumed by what is happening on the screen, where the same boy as always is talking to the robot copy of me. This project takes up all of his attention to the point that he has practically abandoned the dueling table. And today there is a face-to-face soirée at the club.

7:38 p.m. – A couple of minutes more and I will notify him. I calculate that this way at 8:00 p.m. he will start getting ready and thus will arrive on time, not like the other day when I notified him with the correct amount of time to spare and, since he repeatedly shot me down before paying attention to what I was saying, he was late and I was severely reprimanded.

7:40 p.m. – I emit: "Time to get ready for the face-to-face soirée, Doctor."

"What?" He takes his eyes off the monitor for a second and fixes them on Alpha+'s clock. "It's not even eight yet! Maybe rust is slowing you down, you heap of scrap, but not me. Go on, start getting ready, and shut up, I'll be ready when I need to be."

7:42 p.m. – The first round went as expected. I will get out of his sight and in two minutes I will come back.

That oaf has a talent for bad timing, a failure that is difficult to predict and even more difficult to fix. The damn thing had to interrupt just when ROBco was informing Leo of everything he'd collected on his journey around the data banks. Now he'll have to catch up with what they were saying a minute ago and at the same time watch the live images, a tiring task that requires great concentration. This is an area in which he does notice the loss of his faculties, before he could follow up to four recordings at once. And now ... he's not even capable of beating Hug 4'Tune in two battles in a row. Since the monk riddle he hasn't won any ... He needs this kid to hurry up with the prosthesis. It seems that his thought process, caught up in a vertiginous decadence, is a funnel that always leads to the same place.

On the screen that's transmitting recorded images ROBco is saying:

"... Following your instruction that 'a genius is someone who looks and lives beyond their time,' I went to the inter-century adoption clinic." Leo's clueless look makes him add: "*Explanation:* Children who were frozen because they suffered from incurable diseases are now cured and put up for adoption. Exactly what you asked for, right? People from the turn of the century transplanted into the present."

The Doctor, born in the first half of the twenty-first century, is sure everyone from his generation isn't a genius just because they were born then. What drivel. Hoping that the boy won't buy it, he's surprised by the answer:

"Wow, you're a genius! But they must be really little, right? Have they lived long enough to really be considered people of the last century?"

They were doing so well introducing records taken from artists, inventors and truly unique people, and now they're losing their way with this digression into kids' stuff, which was never a more appropriate phrase.

"*Details:* Lately they have been unfreezing fairly old ones, eleven or twelve years old, there is even a girl who is thirteen. *Proposal*: Next week there is a get-together for adopted children; if you authorize it, I will go."

Alpha+ interrupts at a bad time again, and receives another rebuff, while the screen continues showing their wayward dialogue:

"Yes, of course I authorize it. I'll go too."

Having reached the end of his tether, the Doctor interferes without thinking it over, which is against the rules.

"Leo, stop that rubbish right now and get to work."

The kid looks at him, worried, from the live feed screen, and he realizes he was watching him on the recorded screen. It's obvious that he's been back to work for a while. He redirects his ill-timed interference as best he can, and ends up making him walk back his promise to go to the gathering. Instead of running around after snotty little kids with no substance, Leo should introduce the Doctor's own records: he really is a verified genius. He's gone overboard with the bravado, and in the end he has to accept Leo's rather reasonable argument that the idea is to complement his creativity, not duplicate it.

8:00 p.m. – "If you do not let me change your clothes and smarten you up right now, Doctor, you will be late."

For once, Alpha+ has been opportune, showing him to be a busy man who has to cut off the communication—one that he pretends is insignificant but deep down bothers him—in order to attend to more important matters.

As he is used to dressing scruffily around the house, agreeing only to bathe himself and be massaged, but never to be shaved or have his nails cut, the task of getting him ready to go out has become extremely tiresome. Every time he feels less inclined to do it. What's the point anyway? To go around showing off his wrinkled skin? It wouldn't be worth it if it weren't that these face-to-face gatherings are his domain, the last bastion where he can reign happily. In long-distance battles, his brain has started to beat itself into retreat, but face-to-face, his aplomb and his gift for simulation and trickery glow brighter than ever. He's ground everybody down and beaten their will. In betting games he has no rival.

He engineers all manner of strategies in order to undermine the spirit of his opponents. Just today he considered painting a red mark on his forehead to intimidate Hug 4'Tune, by reminding him of his spectacular defeat, but it occurred to him that he would also be lowering himself by awarding it so much importance, and so decided against the idea.

Despite having gotten him into a presentable state in record time and expertly driven him to the club in the aero'car, Alpha+ receives a historic scolding for having arrived seven minutes early. He doesn't stop cursing it and calling it a saboteur during the two detours he makes it perform to waste time, far from the entrance. When has the great master ever arrived before his acolytes? This has never happened to him before. While it could be him losing his flair, this ROB certainly isn't learning anything. Gatew will hear about this.

When he finally gives the order to approach the entrance, it angers him to discover that a line has formed, so he sends Alpha+ to find out what's going on.

9:14 p.m. – "They are asking for a password to enter, Doctor."

"WHAAAT???" His cry is so loud and sustained that everyone becomes aware of his arrival.

Immediately Hug 4'Tune comes over to explain. He says that it's part of the activities they've programmed for today: the speed with which the code is deciphered determines the position that you will occupy at the poker table. It's vengeance for the person who's in penultimate place in the ranking, it was his turn to organize and he's decided to use it to his advantage; he has the rulebook on his side. He shouts out a number from zero to twenty over the speaker system and a number must be called out in response. It's like a raffle: a simple guessing game. It must be something trivial, because the man in charge is the slowest of the group, but up to now only two players have been able to enter.

The Doctor's initial anger has turned into impatience and he cuts Hug off brusquely: he should limit himself to giving the details and forget about the backstory. The boy shows signs of being hurt, but soon pulls himself together and continues. One of the players has had a stroke of luck responding "six" when the number twelve was called; then everyone started halving the numbers, but the door didn't open; until six was called and some fool tried the strategy again answering "three" and, to everyone's surprise, was let through. Next they called seventeen and seven, but there was no luck with the prime numbers; and the last number to be called was one, which was also answered unsuccessfully. There's only one girl in front of them and then it will be their turn.

Just then he hears "nineteen" over the speaker and the girl replies "thirteen." There is no response, apart from a metallic voice from behind them that takes them by surprise.

9:20 p.m. – "If you want to prepare, the next number will be nine."

This causes some commotion, since it's prohibited to accept a ROB's help, and, faced with the Doctor's threatening stare, Alpha+ feels obliged to clarify that the order of the numbers coming from the speaker has nothing to do with the correct response, revealing it has not broken any rules.

Unaware of what's going on around it, the speaker emits an implacable "nine," and the Doctor, surprised by his ROB's correct guess, gives Hug 4'Tune a complicit look just as the boy answers "four" and is invited to go inside. From his apologetic gesture he's not sure whether the strategy of dividing by two and rounding down worked for him by chance or if the son of a bitch had deciphered the code and was playing an act. He shouldn't get worked up: a duel's a duel. Puzzled by what has just happened, the Doctor has let two numbers pass by, and only tunes back in when he hears Alpha+ say:

9:23 p.m. – "The next number will be five."

As fast as he can the Doctor pulls up all the examples he has: twelve/six, six/three, nine/four, and now five/…? Like an epiphany, he sees the five/ superimposed on the nine/, both of them the same length, four letters. Ah, nine/four, that's it, "nine" has four letters, and "six" has three, and "twelve"…the speaker hasn't yet finished pronouncing the word "five" when he replies "four," and they show him inside.

He's so pleased with himself that he doesn't mind giving Alpha+ some of the credit. Having the "five" beforehand allowed him to focus his search, whereas discovering the general pattern would have been more difficult. A great help, and it was smart enough to do it without breaking the rules, throwing him a rope that couldn't strictly be considered a clue. If it carries on evolving like this, maybe it will become the prosthesis he needs, and he will stop depending on that trio of arrogant scatterbrains from the E-Creative project. Taking advantage of the privacy of the decontamination chamber, he congratulates it enthusiastically in order to reinforce that particular behavior as much as possible, while at the same time showing he's interested in the sequence the speaker is following. He's intrigued by the underlying logic, but, above all, by how Alpha+ came to see there was such a logic.

9:28 p.m. – "Mr. 4'Tune said that the organizer was slow, and guaranteeing randomness would be very difficult, it's easier to follow a pattern."

Fucking heap of scrap, now it's a philosopher and everything. With its mechanic reasoning it's got humans down to a tee. It turns out we're the ones who are routine by construction, and they're the only ones capable of being unique. We're always attributing our inspiration to randomness, our most creative solutions, but maybe subjectivity has some inflexible laws that we are all bound to follow. Everyone except the ROBs, of course. If creativity really is a random generator, they are the future, the prosthesis I'm looking for.

"And what was the pattern?"

9:31 p.m. – Warning! If I tell him, he will accuse me of allowing his brain to atrophy by giving him everything on a platter. "It is better for you to discover it yourself. Remember that they were numbers from zero to twenty, and the pattern began: zero, two, twenty, twelve...Do you notice any consistencies?"

"Does it have something to do with the letters like last time?"

9:32 p.m. – "Yes, but different. Should I continue? Three, thirteen, ten, sixteen..."

"Stop, stop! Six...and the next...probably seventeen? If I don't write them all down I don't know how to order them."

9:33 p.m. – "For me, just setting the descending-order option made them come up that way."

Everything is in their favor, the Doctor thinks, while he crosses the threshold into the room where Hug 4'Tune, his wife and Sus Cal'Vin come over to greet him, ever attentive.

"Good evening, Mrs. Poodle," he lets fly as he performs a mock bow in Fi's direction, "your boxer gave me a run for my money tonight."

"Dr. Craft, with the utmost respect"—she's shaking like a leaf, but wants to appear firm—"don't do it again, please."

"Good grief, 4'Tune, I didn't know your companion was so sensitive." He turns away as if the woman were no longer there. "And you, Dr. Cal'Vin, what are you doing here?"

"A service mission, as ever, Doctor."

"Don't tell me you're working for CraftER, here, at this time of night."

"Well..."—she looks around her, as if she thinks someone may be spying—"I can't explain right now."

"I'm not really interested anyway! But since you do work for me, tell me a couple of things about that Leo Mar'10 you interviewed."

"A pleasure, Doctor."

"What kind of relation does he have with kids?"

"He has no children of his own, that goes without saying. The protocol is very strict on that point." She feels Fi's gaze on her and tries to forget about it.

"I mean does he get on well with them, is he drawn to them in any way?"

"We didn't talk about that, but he's so presumptuous and maniacal, I wouldn't put anything past him. I'll look into it, if you want. Do you suspect him of something?"

"That's debatable. Let me know as soon as you find anything out. I hate kids and I wouldn't want them interfering in my project in any way."

"You can't imagine how much I understand." It's getting more and more difficult to ignore Fi's outrage.

"Of course, that's why you have the position you do at the company."

Sus needs to change the subject, it doesn't matter how:

"And the others, Miq 6'Smith, the Picasso girl, do you want any more details? I shouldn't think you'd need anything on him, and her ... Well I did try to talk you out of it; by the way, she didn't even know when Picasso was alive."

"That's why I picked her, she wasn't contaminated by high culture prejudices"—he waves his arm pompously—"like you and I ... and perhaps 4'Tune."

The aforementioned man, who has stoically been putting up with these affronts to his partner, sees the perfect opportunity to intervene:

"We should go inside. The game is about to start."

The Doctor waves goodbye to Sus, like someone who is off to carry out an important mission, chest puffed out, and tells himself he's getting taller with every step, while Hug, his loyal squire, flashes the women the look of a lamb being sent to slaughter.

Fi, doubly irritated by the rudeness of the Doctor and the servitude of her husband, is quick to take her bad mood out on Sus:

"How can you be so cold and turn your back on Xis? 'You can't imagine how much I understand,'" she spits at her, imitating Sus's voice. "Your brilliant career is built on sand. Hasn't it ever occurred to you that you could be found out?"

"After all this time ... no, to be honest. It's easier to hide an implausible feat, no matter how large, than a plausible trifle. The Doctor could never even conceive of me having a daughter."

"I wouldn't be able to hide it."

"Of course you wouldn't, but you don't work either, so you don't know what it's like."

When the conversation comes to an end, Fi concludes that, instead of lending her her daughter to go to parties and celebrations, Sus would be better off giving her the girl to keep.

16

Mom, can you see me? I hope so; it took me forever to find this spot. You can't see the sky from anywhere in the house. Do you think I'm exaggerating? No, of course not, you must have seen it. A huge ceiling full of holes ... is that what you see now? Well underneath it there's another, the one that belongs to the platform where I am now, and at home there's a third. That's why I can't see the sky. It's really hard to line up three holes, four if you count the one in the ring, although that one's easy to do. I've tried it from all the windows, crouching down, climbing up on things, looking out sideways, but it doesn't work. The only thing I haven't been able to try is leaning out. The openings—that's what they call them here—aren't really windows: they don't have a frame and they can't be opened; they're bits of see-through wall in any old shape. They kind of look like portholes.

I've been lucky. When I came outside, I didn't think I'd be able to line up one of the dome's holes from our own platform. It's so close I'll be able to come all the time. How long's it been since we last spoke? Since that day Xis caught me doing it on the school aero'bus. I had such a hard time explaining to her that I wasn't talking to myself even though you weren't replying. At least when she saw the ring it convinced her I hadn't made it all up. She liked it so much she wanted me to give it to her. They obviously don't make them out of gold anymore, and less still with these beautiful designs. She tried it on and said it gave her the shivers. I find it more like a pleasant tickle and, since she said that, sometimes I brush it up and down my arm and imagine it's your hand that's stroking me. That day, however, I was afraid she wouldn't want to give it back. She always wants to keep everything for herself: one day at school she asked me to give her my hair, thinking it was a wig, and, to talk her out of it, I had to let her pull it. Can you imagine? It was horrible, but what else could I do if I didn't want to lose my only friend? You told me that in the letter: "it's good for you to go out and make friends," remember? You see, I already know it by heart, and I did as you said, Mom. Are you happy? It changed Xis' mood, she didn't smile, because she never smiles, but at least she wasn't angry anymore. She's almost always angry, you know? I think she must be going through some tough times. She's strange, I would like her to trust me and tell me what's going on with her.

When I got home Lu yelled at me for letting my hair down, I have no idea how she found out, she must be spying on me all day. That's why I like this place, here behind the aero'car is a blind spot on the surveillance system, I hope, and ROBbie has promised that he'll make it look like I'm with him and, if she starts looking for me, he'll warn me.

Oh, Mom, I'd love to tell you everything like when I came back from summer camp, but so many things happen to me here, and it's all so hectic I'm not sure I'd be able to explain it all to you. They've hired a home tutor for me. I can tell you're surprised and I can guess what you're thinking: "For you? But you've always done so well in school." But, you know what? There are no subjects nowadays, they just teach you to use EDU-sys and to behave. You don't have to memorize anything, like before in geography and history, and you don't have to learn formulas either, like we did in math, the ROBs do that. It's like they're teaching us to play, first on our own with the computer and then in a group in the socialization room. Because there's no playground, can you believe that? And no break time. They don't sleep much either, I think I told you about that, for the first few days I was practically sleepwalking, until Lu worked out that I needed to sleep more.

She does everything she can for me, and I try to make sure she's happy, even though I don't always understand her. She's hired a tutor to help me but she keeps interrupting our sessions and getting in the way, as if us working annoys her. Just yesterday she pulled me out of the session to ask me if I was okay. I don't know what she's afraid of. It must be what Silvana was saying about being an emotional masseuse ... Don't be alarmed, she's never done anything weird to me, or used any of the sophisticated devices they have now, actually she's against those ... she doesn't even have a ROB. Maybe that's why Lu doesn't trust her.

It also surprises me a bit that we don't do any homework and we haven't connected to EDUsys, we just chat. She asks even more questions than you, Mom, but I don't mind: I like talking about all of you, what we used to do and which differences hit me the hardest. She says that to be able to help me the first thing she needs is to know what I'm like and how I feel. According to her, my problem comes from that, from the fact that I react differently than kids that are around these days, and that's why EDUsys has problems programming my education. I was really pleased she said it was the robot that had problems, not me. She told Lu that too, and she almost fired her.

She was already a bit fed up because, right from the start, Silvana wouldn't let ROBbie attend the sessions. To tell the truth, I was a bit disappointed. I've become so used to always having it close by that I feel strange when it's not around. She wouldn't budge, though: I need to get my sense of identity back—that word's really stuck with me—it has to be me that accepts new stuff, not the new stuff that beats me into submission, you might say, and cancels me out. I wasn't sure if ROBbie annoyed her because he was watching us, or because he was a new thing that would distract me when I was supposed to be remembering my previous life. That's what she makes me do: remember.

When I was talking about good moments, I told her that in the summer we used to go to Gurb, to Grandma's house, and we would spend the day outside: helping the workers harvest potatoes, feeding the chickens and the rabbits, riding our bikes and, sometimes, swimming in the neighbors' pool. She was really taken aback by the fact that year after year I met up with a group of friends that I didn't even speak to during the winter. It was hard for her to understand that we could reconnect after so many months of everyone going their separate ways. It's like we lived two lives, she told me, and you put one on hold so you could start the other. At that point in our conversation I still didn't know that school is open all year round, and that I wouldn't have summer vacation. I was so close to crying. Silvana realized right away and tried to cheer me up. If I wanted a change of scenery, she would convince Lu to let us do the next session where she lives, which apparently is really different from here. She reminds me of you a bit, Mom, she listens to me properly and looks me in the eye, like she's trying to reassure herself that I'm not getting distracted and that her words find their way into my head. And they really do! Since she told me about it, I can't stop trying to imagine what her house must be like!

It does worry me a bit, you know? It seems like she's using tactics to get me to talk. You were very good at that too, but after talking to you I felt more relaxed. Not with her. I'm always afraid I've said too much. And who knows if Lu, who must be spying on us the whole time, also thinks I tell her too many secrets and that's why she pulls me out of the sessions. It is true that I hardly know Silvana and that she, on the other hand, knows a lot about me.

Do you think she likes me? Considering she's invited me to her house, I suppose she does, but then I wonder if it's just part of her job... she seems very professional. She asked me to describe you. Yes, you. Dad and Grandma too, but it was obvious she was mostly interested in you. I really sang your praises, you can imagine, and, while I was speaking, my eyes filled with tears. "You love your mom a lot, don't you?" she said, and she stroked my cheek with the back of her hand to dry my tears. It's the only time she's ever touched me. I told her that you were really strong and that I'd only seen you cry once, when Grandpa died. You don't mind do you? That I talked about you? That was the day I felt worst of all, it was as if by talking to her about it I had betrayed you somehow, although I didn't say anything bad. In the evening I was worried but I didn't know why. Now I'm happy I've told you about it.

Bye, Mom, I wouldn't want Lu to find this hiding place. I'll come every day. See you tomorrow. Love you.

17

Operation: autonomous. I am only authorized to contact Leo about extreme cases, and then I must avoid all reference to the inter-century adoption meeting; otherwise I will be penalized. *Motive*: My PROP will be at risk if Dr. Craft finds out where I am. *Task*: I have to search for the most creative adoptees. *Restriction:* I need to attract as little attention as possible.

Status: bifurcation. A screen displays forty-nine activities, many of which are simultaneous. I am unaware of the meaning of some terms: "post-adoptive support," I can work that one out; "open adoption," must be defined in opposition to closed adoption, the same as operations that once they are closed, are irreversible. I *suppose*, then, that open means that the child can be returned, but it would be better to find out for sure. I *activate* the word and read: "in open adoption, the three vertexes of the triangle (biological parents, child, and adoptive parents) are identified, they know each other and attempt to establish links between the three." I was wrong then, I must widen the definition of "open." I *consider* whether this type of triangle would interest us. Probably not: in order to create a strong cultural shock, the child must have been frozen for as long as possible and, therefore, the parents must be dead. Or maybe it does: although they might be dead, it's best that the child remembers their previous life as much as possible to notice the contrast and, therefore, it is necessary that they have been frozen as older children and retain a strong emotional link with the people that were their parents. I *inhibit* the warning to ask for direction from my PROP and decide to assign priorities according to the fixed criteria I have.

Mode: define strategy. I ponder the four criteria set out by Leo: (1) older children, (2) recent adoption, (3) frozen for an extended period, and (4) relation with creative tasks, and by combining them I obtain a priority rating for each of the programmed activities. I *compose* an optimum itinerary according to this priority system. Under normal conditions, I would refine the result with my PROP, but, due to a lack of additional information, I accept it as is.

Process: execution of itinerary. First destination: theater workshop for children aged between ten and thirteen who have been unfrozen for a maximum of two years. I localize area D17 and go in, trying to remain undetected among the twenty or so ROBs that are already there. One member of the organization is giving instructions to six boys and four girls, closely watched by eight hypothetical mothers and a couple of fathers. I *observe* the children one by one looking for subjects of interest, and I realize that each one has two holographic numbers on their back. I assume that the first is their biological age and the second, which in no case is larger than two, must be the time elapsed since they were adopted.

Subprocess: identification of the Most Valuable Subject. I *save* the visible numbers, and I move around to see the other children's backs, save them and move around again. The instructor makes the task more difficult by constantly getting between the children and myself. I move up and down and turn from side to side to get a better view. What a waste of energy. *Warning*: useless thought. *Action:* I inhibit this indicator now that I have unlimited resources provided by CraftER. I *continue* collecting numbers. I am not streamlining my movements to achieve my objective. Some of the ROBs start to look and, then, some parents do too. *Alarm:* it is necessary to remain undetected. I stop what I am doing.

Status: tactical pause until attention is distracted away from me and is refocused on the performance. I make the most of this time to record what is happening on stage: they are all sitting down in rows while girl 12/1.8 scribbles virtual doodles on the wall. In order to interpret it, I need a context, a pattern. I *recover* the initial instructions of the organizer: "You will begin by acting out a scene from a school in the past."

Suddenly the one who is playing the teacher turns to a student in the front row and asks them a question. After that she makes another one stand up, and another, until someone makes a high-pitched noise, and they all stand up at the same time and, pushing each other and laughing, they head to the back of the room in a big group. I cannot identify any of what happens next, their movements become chaotic. I look around and realize that no one is looking at me.

Action: I take up the pending subprocess. Three of the identifiers I was lacking are now visible. I am only missing one. I successfully rush to get a snapshot of the hidden back. Data collection complete. I *calculate* the MVS: 13/0.4, and localize the subject. *Save:* white female, 5'4", slim build, extra-long brown hair, light skin covered in unknown marks, I include photo... *Interruption:* A robot with the name ROBbie is coming toward me.

"Friendly greetings, ROBco. Why are you observing that girl? I inform you that she is my PROP. Which is yours?"

Strategy: Admitting that mine is not present will lengthen the interrogation, but I stick to the truth. It is the only way I can possibly obtain the MVS's encephalic records.

As both ROBs are recent models, the communication flows easily, with no need for any editors or converters, and ROBco can quickly formulate a petition for the girl to collaborate in a high-tech project.

"*Denied*: It requires consent from the mother and she does not speak with unknown ROBs."

I *perform* a sweep of the room and *save* the image of the woman, in case it can be of use to Leo later, while I *emit:* "But you can speak to her."

"Not until we get home. With the tutor present, with whom they are talking now, I am not permitted to approach any of them."

Warning: dead end. *Action:* Move on to localizing the next MVS, 12/1.8. *Precaution:* save prior context. I take an image of the tutor as well, whom I catch intently focused on the gesture that 13/0.4 is making at her from the stage. Unexpectedly, the girl fixes her eyes on me, and consequently so does the woman. *Alarm:* possible malfunction of the camera, leaving a physical trace. *Consequence:* added difficulty when trying to remain undetected. *Emergency*: I stop all movement and turn off the camera.

Status: peripheral hibernation.

Unexpected event: urgent request from PROP. I *open* the internal line: "ROBco, I want you to auto-diagnose right now. It must be a serious problem for you to turn yourself off at a critical moment." *Basic test:* correct functionality. *Explanation*: I have entered hibernation as a precaution and the situation is not critical. "You're letting a unique subject get away, thirteen years old and only unfrozen for four months, and you're saying it's not critical?" *Information:* 13/0.4 is inaccessible, and 12/1.8 and similar are plentiful. "Exactly. Learn, learn from your data: there is one, and then two years later all the others, that makes it a singular subject. I order you to apply all your available resources and energy reserves to getting in contact with her." I *repeat:* the subject is inaccessible. If you wind back, you will hear the dissuasive reasoning of her ROB. "Okay, you can't talk to the mother, but I can. Why didn't you notify me?" *Reminder*: I would have been exposing you to punishment by Dr. Craft. "In an extreme case like this one, you have to take risks. Learn that too. I'm on my way over. Whatever you do don't lose sight of the girl." *Notification:* the punishment may be ... *Warning:* useless communication, there is no receptor.

***

On the way over Leo puts his strategy together. Between ROBco and the girl's ROB, to which it seems to be well attuned, they'll have to distract the tutor while he talks to the mother. If, as he suspects, she is anti-techno, she will only be a nuisance. He won't explain the project to her in its current form, but as a collaboration with CraftER and, with the snippet he's going to explain to her, the woman will see the path to fame. The stereotypical figure and challenging stance he's seen in the images show her to be someone who is vulnerable to this temptation. She has the same ambition as Bet in her eyes. He'll win her over.

When he gets there, however, the situation has altered somewhat. The mother isn't there, and the girl and her tutor have thrown themselves into participating in a dance circle. When he asks ROBco for an explanation, he realizes that he himself had ordered the robot to follow the wrong subject. Luckily, a very helpful ROBbie tells him that Lu, as the mother is called, has gone to the Adoptive Families Resource Center stand to find out what's available. She'll probably be back soon.

Despite loitering discretely in a corner of the room, Leo fails to go unnoticed. It's impossible to do in a place where only children, ROBs and women of a certain age are present.

"Look, Silvana," Celia whispers, "that must be the PROP of the robot that's been following us around. Do you think he wants to talk to us?"

"I don't know. Do you know whose dad he is? He looks very young to have adopted a child."

"He just came in, didn't you notice?"

"I see *you* did." She looks Celia in the eye and smiles. "You can't have seen many guys that age, right?"

"Now that you mention it, he's the first." She stops to think this over. "It's odd that I didn't notice."

Silvana also stops, the music is fading out.

"It's one of the drawbacks that comes with living with the pro-technos. Your mother would say I'm preaching, but the fact is that here each generation lives in their own little world. When you come to the ComU you'll see it's really quite different."

The circle breaks up in no time at all and, before everyone hurries off to the next activity, the organizer reminds them that the identification of all the participants will be sent to their houses in case they want to get in contact with each other later on.

Standing quietly in his corner, Leo pretends to be distracted without missing anything that's going on. The little girl keeps looking at him out of the corner of her eye, maybe he should go and talk to her without the mother's permission. She seems bright enough. Every time he's about to go over, however, the sight of her tutor puts him off. He wouldn't know how to act faced with a woman who must hate anything that isn't material. The good-looking anti-techno tend to be like that. When they're still young and sexy they reject technology. Of course, if all of a sudden plastic surgery and thought enhancements were abolished they would have no rivals. And him? What could he offer, if his exceptional command of technological tools were no longer useful? The woman seems to have read his mind, because she's giving him an arrogant look that he makes an effort not to pick up on. In spite of her age, he can't deny she's attractive.

Although she's aware Silvana won't like it, Celia goes over to ROBbie, since she'd seen him talking to the young man's ROB earlier, to ask what he knows about him. Not a lot, to be honest: he was interested in some information about her and he's waiting for Lu to ask her permission to collect it.

"Will you come with me to ask him what he wants?" Curious and a little on edge, Celia wants Silvana to come with her.

"Of course, the one time you have an opportunity to meet someone different I'm not going to deprive you of it!"

When he sees them coming, Leo stands up respectfully and puts some distance between him and his ROBco, so that the robot's presence won't make them uncomfortable.

"Hello, my name's Celia." She holds out her hand and is surprised at the ease with which she takes the initiative. "ROBbie told me you want some information."

It's obvious that the young man isn't used to this kind of greeting, but he quickly adapts and shyly takes the girl's warm little hand.

"Yes, how good of you to come and see me." When he smiles, his eyes light up. "My name is Leo." He looks at Silvana, but she deliberately hangs back. She'll be watching me like a hawk, he thinks, I'd better be careful what I say. "I study creativity and, from what I've heard about you, I think you might be able to help me."

Struggling to contain her excitement, Celia rocks back on her feet slightly, looking for her tutor's blessing, but she remains unreadable.

"If I can, I will." Finally she's grasped that it's better not to be too forward. "Are you a psychologist?"

"No, I'm a bioengineer." The girl's clueless expression prompts him to explain himself. "I work with machines that process human signals. Do you understand?"

"Of course, hospitals are full of those machines."

Suddenly he realizes that all these adopted children were once terminally ill, and that submitting them—submitting this little girl—to his experiments might be a kind of torture. He is embarrassed by this lapse, and it makes him wonder whether, inside his cubicle at CraftER, he had been aware of and accepted these consequences as part of the project. The timeout device is making him feel like he's going crazy.

"Are you afraid of them?"

"The machines?" She looks at him skeptically. "No, I really like them." She remembers how she enjoyed it when her dad explained them to her. "But no one can tell me how they work anymore. Can you?"

"Some of them, of course." So many doubts and it turns out it'll work in his favor after all.

"For example, do you know how robots pick up far-away conversations when there's a lot of noise around?"

"They've got powerful antennas that emit and receive electromagnetic waves on a very narrow channel, and, if they need it, they have a source separation program that helps them."

"Wow, cool!" She's used an expression from before, that sounds old-fashioned to her now, and she turns to Silvana to gage her reaction.

"Who do you work for?" Finally, the highly anticipated intervention of the tutor.

"Don't get me wrong, I'm doing an investigation. For now, it's nothing more than that."

"There must be some company behind it," she announces severely. "Which one is it?"

"Okay"—hiding information can only be counterproductive—"CraftER."

"I knew it! The ones who make more and more intelligent robots for ever-stupider humans ... and now you want to destroy our creativity by passing it on to robots?"

Celia gives her a troubled look, she'd never imagined her tutor could be so aggressive. Especially not with someone so friendly. Leo, on the other hand, is made bolder by the turn in the conversation, as he has some great arguments:

"No, no, it's quite the opposite. It's about strengthening human creativity, making a kind of devil's advocate that spurs it on."

This is followed by a moment of silence, during which Silvana weighs whether the person she's talking to is shameless or naive, and, lacking the information to make a decision, goes for a bit of both:

"You lot are always nit-picking: you're not trying to replace anyone, just broaden their abilities, that's why you rush to use euphemisms, like assistant or helper, instead of saying executor or usurper, which is what they'll end up being."

She's regurgitated the standard speech and even Celia realizes that Leo hasn't mentioned assistants but advocates.

"Okay, let's take this bit by bit. I can't answer for everything the company has done, but I can assure you that my research ..."

They all turn toward Lu, who has arrived feeling very overwhelmed and, upon seeing the boy, has stopped some distance away to recompose her expression the best she can.

"The Educational Media Exhibition started a while ago, but I can see you've been busy."

"I'm sorry, please excuse me." Leo was quick to react and in two strides he's standing next to her. "Could I talk to you for a second?"

"Yes, of course, I'll be right with you." And, to Silvana: "You can head over, I'll catch up in a second."

"I'd rather wait here." Celia says this in a sensible tone of voice, slowly crossing her arms to affirm her position and delay her possible departure.

"Oh, sweetie, don't be so direct. You sound like a robot talking like that."

Despite Celia's insistence on staying, the two women are for once in agreement that the EME is the most interesting activity of the day and they can't miss it. Between the two of them they've gotten the better of her, she wasn't counting on them teaming up, and even less on their alliance resulting in such a resounding defeat. She only hopes that Leo's proposition convinces Lu... and that Silvana doesn't get in the way.

18

Since he's become accustomed to the quiet isolation of his cubicle, Leo's head is spinning when he returns from the event. All he can see is a big mess of images on top of one another, and his inability to order them is really starting to get to him. At just thirteen the girl had gotten the upper hand in their conversation. She's so curious about everything, she seemed to soak up his words with her eyes. And to think that ROBco would have let her get away! Thank goodness he'd thought to supervise it; despite updating it to a cutting-edge model, he still can't trust it. Even with the neuroaccelerator it learns so slowly that, at this rate, it'll take years to get the Alpha+ type assistant he'd expected. He's starting to understand the rise in popularity of second-hand ROBs. Depending on the circumstances, he may also prefer a pre-trained robot, even if it were attuned to someone else's peculiarities. Dr. Craft's, for example. Now he just has to tell him that, by disobeying him, he's made a great discovery that will advance the project, and therefore he needs a little girl to come into CraftER so he can record her encephalic records. That's if that anti-techno halfwit doesn't brainwash her mother. He reckons he managed to convince her, and Celia—he'll never forget that name—is a great secret weapon to have on his side. The determination she showed when she insisted on staying! Strong-willed, interested and stubborn, she has all the characteristic traits of creative types. And cultural shock is guaranteed, he couldn't ask for more. He'll set up an interview with the Doctor. He's sure he'll see why this unique person is so useful for the project, and he'll come around in the end.

When he tries to call, Alpha+ blocks the connection, claiming the Doctor has just lost a dueling match and is in a very bad mood. If he wants to leave a message, the robot can pass it on at a good moment. This is the kind of assistant he wants, oh yes, capable of making decisions that benefit all parties. When he reveals his intentions, though, the robot kindly informs him that the Doctor is not at all keen on children and, as far as it knows, he's never interviewed one. Leo wasn't counting on that. He decides that the best course of action is not to tell him anything and go for broke with the presentation of the demo. If it's convincing enough, maybe he'll be prepared to make an exception.

He dedicates himself day and night to perfecting the prototype and, by really pushing the robot to its limit, manages to get it to do the dirty work of entering the data and running an exhaustive battery of tests to check that it's working properly. He's so absorbed in the project that it seems like everything is related to it, and everything gives him new ideas. Even his expertise in debugging programs leads him to an unexpected goldmine when he realizes that, when applying these tools to one person's behavior, they will detect actions susceptible to improvement or innovation, which would permit the suggestion of more creative variants of the same behavior. His modular, open and flexible algebra allows him to incorporate this strategy and any other almost in real time, without great effort needed in coding. That's the advantage of a powerful formalism.

Instead of coming to the cubicle as expected on the agreed date, the Doctor simply appears on the control monitor to give Leo instructions: he wants him to send ROBco with the prototype. Leo, who has forgone hours of sleep and is nervous about his gamble, just about manages to recover from the surprise in time to tell him that he's prepared a presentation.

"You want to torture me with superfluous chitchat as well? What's wrong with you engineers these days, the only thing you know how to produce are fatuous words! I warned you..."

"No, Doctor, I'm convinced you'll like the prototype, I've developed it just as you wanted, structured in layers, but I should explain..."

"Products should speak for themselves."

"I agree ... if they're finished; but at the moment I've only completed the first layer, the reactive one, that, when requested by the user, tries to provoke original solutions in a specific field." He tries to milk the maximum amount of information out of the minimum amount of words. "I want you to know that I'm working on the proactive layer, which constantly monitors behavior in order to emit suggestions, and on top of that will go the intrusive layer, with direct access to brain signals to open much deeper paths of innovation."

"As you can see I decide what I want and don't want to listen to. You may continue."

Spurred on by this comment, Leo gets carried away and opens some graphics from the presentation, gesticulates, gets caught up in the details, without realizing that the eyebrows are starting to arch perilously while he continues enumerating inspiration strategies, mechanisms to focus human attention, and even permits himself the frivolity of trying to flatter the Doctor, citing reliable sources, by stating that the key factor in being original is wanting to be, as Pasteur said. That's when, in turning around to reap the fruits of his audacity, he is shocked to find that the image of his only listener has disappeared.

Furious with himself for having been so inept—"you have to be stupid to forget about the most important thing"—he starts to curse himself out loud, aborts a first kick, punches at nothing, and ends up fighting with the air.

ROBco watches him inexpressively, without attempting to intervene, until all of a sudden it lets out: "Alpha+ is asking for the prototype," and, since no reply is forthcoming, repeats it, and repeats it again.

At first Leo doesn't even hear it, then he ignores it, and in the end, exasperated, he tells it to shut up. He can't understand what has just happened, he's never experienced such an uncontrolled reaction and he looks down at his own body in amazement. Maybe the reclusion is driving him mad. The gloomiest predictions of his former colleagues about inventors being wrung dry by an unscrupulous tyrant come back to him with force, exaggerated by the Doctor's phantasmagorical appearance and even worse, his unexpected disappearance. Those words "you want to torture me as well?" are still ringing in his ears. At the time they'd given him the shivers, and the fate of those who came before him worries him more than ever.

When ROBco reminds him again, he jumps to the control panel—the situation couldn't be made any worse by the Doctor receiving the prototype—and extracts a self-contained version of the system, ready to be installed in a ROB; once it's been copied onto one of CraftER's confidential storage devices, he hands it over to the robot to deliver and, while it's at it, leave him to think in peace.

Now that he's on his own, he automatically leaves the cubicle so he can contact Bet. He's always done this upon finishing a project. What's not so common is for her to respond right away. After exchanging only a few words, she delivers the verdict:

"I see you have your typical post-demo depression."

It's true that this has happened on other occasions after throwing himself body and soul into a job; when he finishes, he feels empty and doesn't know what to do with himself. But this time it's different, he claims. Since he doesn't dare to share all his fears with her, he gets tied up in knots insinuating doubts about his abilities, and she interprets this as terror at the thought of receiving a critical evaluation.

"No, it's nothing like that. If I'm sure of one thing, it's that he'll like the prototype."

What are you afraid of then, the next step?" He's got her pretty confused.

"Kind of, yeah."

By answering with half-truths he starts to spin a web around himself, and his fear of the Doctor coalesces around the fact that, certainly, he will refuse an entry permit to someone who is key to the advancement of the prosthesis. Therefore the project will fail, they'll fire him without a second thought, and, on top of all that, a promising prototype will be snatched away from him.

"You mean it's a premeditated plan? Who is this key person?"

He finds himself deep in conversation about the get-together, how he'd turned up unplanned so as not to lose as singular a person as Celia, about the qualitative leap that she could bring to the project.

"She's not like you or me, believe me: she reacts, she moves, she behaves in a different way. You should see her."

His tone, which had been defeated, is now practically enthusiastic, and Bet is quick to accuse:

"All that back-and-forth to get to this? You'll engineer something to get permission from the Doctor, I'm sure of it, and you'll have your little girlfriend, sorry, your work material, available soon."

She hangs up and, as much as Leo tries to reestablish the connection, unlike before, he comes up against all the intermediaries imaginable.

When ROBco gets back, Leo's mind is still occupied with Celia and the tactics he could use to get permission from the Doctor. The most direct way would be for him to like the prototype so much that he gave him carte blanche. But, even if this were the case, it could take a while, as the installment of the prototype in Alpha+ has been postponed until the president resolves some other affairs.

Leo is lost in thought: what if he were to risk it and request permission through the emergency line, claiming it's essential for the next layer, otherwise the project will have to be halted? Maybe in all the hurry he would be able to leave out the girl's age and once the interview has been agreed to ... Or he could hide it among a large group of people he wants to record, then the Doctor wouldn't have to interview them one by one. This seems like a viable option, in spite of the extra work it would imply.

"Let's see, ROBco, how many people have you identified that we would have to record here, within CraftER?"

"*Recap*: with certainty, only Celia, the others have been discarded. *Question*: What is the objective?"

"To get permission from the Doctor for these people to enter CraftER."

"*Suggestion:* Consider whether this is the final objective or a way of achieving it."

Leo raises his head, surprised, he can't remember ROBco ever having taken the initiative to make suggestions. Maybe the neurolearning is finally starting to make a difference.

"Okay, the objective is to obtain permission for Celia; to place her in a group with other people is just a way of achieving that."

"I *insist:* Is getting the Doctor's permission your ultimate objective?"

"I'm glad you're trying new things, but do you have to repeat yourself?"

"*Explanation:* Superfluous restrictions often get in the way, one must discern which are essential and which can be relaxed." Leo has fallen silent. "*Example*: NASA spent years unsuccessfully searching for a metal that could resist high temperatures for its rockets, for when they reenter the atmosphere. Until someone realized that the objective was not finding this metal, but protecting the astronauts during reentry. The solution was a material with the opposite characteristics of those they were looking for, which burned very slowly, thereby keeping the heat far away from the actual vehicle."

The story sounds familiar. It was in one of the creative-solution dossiers he'd decided to incorporate into the system. Could it be that ROBco had memorized some of the examples while he was introducing the data? In any case it'd hit the nail on the head.

"Okay, the astronaut is Celia. The only thing I want is for her to enter the CraftER atmosphere safe and healthy."

"*Incomprehensible*: Do I have to change the context to interpret this?"

"Forget about it. Let's say I already know the true objective, now what do I do?"

"*Advice:* Look for alternative ways to achieve it. *Options*: Try a drastic change of perspective, or start by disrupting the failed solution a little: change the type of material, the identity of the people, each of the elements that feature in the scenario."

Little by little Leo's suspicion is being confirmed and, although he could try to prove it right now, he refrains so as not to give the game away. He has more than enough reasons to want to go along with this step-by-step, and is curious as to how it will unravel; who knows, it might be useful. One rather drastic solution he's come up with would be to sneak Celia into the building, but he envisions a heap of problems, so he decides to explore the first, more conservative option. There are two elements in the failed solution: the permission and the Doctor. If it's not permission, what could it be instead? And if not the Doctor, who? There's no need to work through any more deductions, or go over any mental lists, because it hits him in a sudden, fully formed flash: Mr. Gatew. He opens his contract with CraftER and reads it over carefully, like a man possessed: the requirement to report exclusively to the Doctor is limited to technical affairs, company protocols should be followed when dealing with administrative issues.

He must be stupid, the most obvious solution hadn't occurred to him! He reads up on the regulations, fills out the corresponding form and sends it to management. He even affords himself the luxury of marking it as urgent.

With the request out of the way, he moves on to ROBco and, with the excuse of having to check that his circuits are working correctly, he opens its side panel and verifies that, indeed, it's been installed with the prototype meant for Alpha+. He's not at all bothered about benefiting from it himself, but he does feel he's been tricked and isn't sure what to think. He'll have to wait a little longer for the Doctor to evaluate his work. What unsettles him most is that the action might have been premeditated—that'd be why the Doctor had given him the same ROB model as his—and also that now it's been done, he hasn't bothered to hide the plan from him. What is he up to? Maybe he wants to avoid taking the risk of installing the prototype in his own ROB and experimenting with it himself, especially since, according to him, there are several prototypes. Of course, it's a way of testing them in parallel. Each inventor, spurred on by their creation, will have their capacity multiplied and will also see their shortcomings exacerbated. What an idea. This way a really positive spiral is set in motion, where differences are magnified, and, in the end, the evaluation will be obvious. Very clever, yes indeed; yeah, maybe he's a son of a bitch, but he's got plenty of wit. Leo just hopes it's him who's leading the field, otherwise his chances of winning will be slim. Luckily, the exchange he's just had with ROBco is a good sign.

19

To Celia this particular school day feels longer and more boring than usual. It seems the day will never end, and the time that Lu is coming to pick her up isn't getting any closer.

For hours she's resisted the temptation to tell Xis where she's going, but in the end she's succumbed, and her friend's revelation that CraftER is the company her mother works at has really surprised her. What a mistake! Since she found out she hasn't stopped begging Celia to pay close attention to everything so she can tell her about it later; she's never been allowed to go. As far as she knows, the building is immense, and it'll be difficult to find her, but, just in case, she told her excitedly, her mother is blonde, not very tall, this morning she was dressed in red, and her name is Sus. This has made Celia even more nervous. She doesn't want anyone muscling in on her adventure.

She can see herself walking along beside that good-looking engineer, past a series of prodigious machines, half-finished inventions, mysteries to be uncovered ... the physics of miracles, as her father would say, and Leo, perfectly kind, will show her first one, then another, he'll explain what she has to do, answer her questions, until the crucial moment when he invites her to sit down—or maybe she'll have to lie down?—so he can set up the devices, or put a helmet on her, or whatever. All his attention will be focused on her. She'll feel so important once she's contributed to technological progress!—her father would be proud—and what luck that the reliable pair of hands won't belong to a disheveled, crazy-faced old man like the experts in her science book, but those of an attentive and slightly shy young man she is so desperate to see again.

EDUsys's insistent demands bring her back down to earth about ten minutes before Lu is due to arrive. Finally the time has passed, but after so much whispering to her friend and creating her own distractions, she's been reprimanded so many times that she fears her teacher might make the most of Lu's visit to share a few of his observations. And she ends up underestimating him, as on top of the usual accusations of a lack of interest in learning and a strong resistance to letting go of old-fashioned and now pointless knowledge, he finishes with a threat: the school won't add even one more day to the period of adaptation than, by law, it's obligated to provide.

Plunged into a tense silence, the journey in the aero'car would have been impossible for the two of them had it not been so short. ROBul is an expert pilot and in seconds they're hovering in front of an imposing building, that to Celia looks like a giant, golden pinecone, planted stem-side down with its scales opened up and some pine nuts still attached. It's nice to look at, even though so much brightness is a bit blinding. As they get closer, she realizes that the pine nuts are aero'cars and she excitedly imagines the maneuvers they'll have to perform to get into one of the empty scales. But ultimately she's disappointed, since the craft immediately heads down and lands on CraftER's main platform, on the ground floor, where Leo has arranged for them to go through security. A mobile beacon guides them to their assigned opening and, once the aero'car has been parked, clear and punctual instructions are received: visitors must wait until their contact from within the company comes to meet them; under no circumstances may the ROBs abandon the vehicle.

Celia's expectations are more than satisfied by the appearance of the engineer. As if by magic, the opening becomes traversable and he crosses it as naturally as George Clooney coming out of a spaceship in that film she used to like. She watches him stride purposefully toward the aero'car and it looks like he's greeted them with a nod of the head, but it's so slight she's not sure it really happened. When they get out of the aero'car and they're standing side by side, she finds herself reliving the feeling she experienced the other day, and stretches as much as she can to make up for the height difference. She almost misses the first words he exchanges with Lu as she's distracted by a brief glance she perceives to have been directed at her legs, and after looking down to find out why, she looks up and identifies another glance, this time at her right hand, she's sure of it.

It takes her a while to understand that the young man is waiting to see if she's going to offer him her hand like the other day and, caught up in her own indecision, she feels a pleasant tickling sensation in her stomach. She's not sure which of them has taken the initiative but she finds herself shaking a hand that holds hers weakly, perhaps because he's afraid of hurting her.

"That's how you greeted each other in the past, right?" Leo is trying to justify the gesture for Lu, but Celia, touched, sees it as an apology for his lack of expertise.

Introductions finished, he puts his hand into a cavity next to the opening—the same hand that a few moments ago was squeezing her own—and, once he has been granted free passage, a disembodied voice announces that management has awarded him credits for two visitors. They must go through the entry control one by one. They have to put their right hand in the cavity so that it can be registered in CraftER's exclusive system and, when they cross the threshold, their identity chip will be read. The association of the registration and the identity chip will give them access to the company's communal areas for the duration of the agreed period. In line with current legislation, they're notified that a record of their movements will be monitored and saved.

As she's used to going through similar controls, Lu has no problem carrying out the procedure. The problem comes when Celia, after putting her hand into the cavity, crosses the threshold and the voice booms: "Protected identity, the name Leo-1 will be assigned as proof of identity of the person responsible." The most surprised is, weirdly, Leo himself, who looks from one of them to the other as if he were expecting an explanation. When he finally reacts and puts the question to the system, he discovers that CraftER, like most companies, only has permission to identify subjects of working age; the actions of a minor will be attributed to the worker who has requested their entry, and he will answer for their actions before the law as if they were his own.

Celia doesn't need to understand all the terminology as she understands perfectly well the meaning of Leo-1 and responsible. She's excited to see herself turned into a kind of double of the engineer, and his doubts, his worried expression when he receives the machine's explanation, offend her. She'd never do anything that might get him in trouble.

Luckily, his eyes don't take long to smile again and he invites them to head on down a kind of corridor dotted with openings on both sides. Celia immediately reverts to being thrilled, and is dying to ask questions. She's intrigued by the fact that they've taken handprints, like the police used to do, she'd imagined they'd have a more sophisticated system.

"So what did you think it would be like?" Leo is pleased to rediscover this habit of noticing everything that had so surprised him last time.

"I don't know, in movies they scan your eyes or your chip ... is that not enough to identify someone?"

"I'm not familiar with the details, none of the users are, otherwise it would be pointless. But you can be certain that, as well as fingerprints, it codifies the distribution of pressure, how long you leave your hand in there, and it's possible that they analyze the DNA in your dead skin cells. It's an exclusive signature that no one else could reproduce because no one knows exactly how it's measured. A chip can always be recoded, and as a last resort, transplanted; the hand, however ..."

"Eyes can't be transplanted either."

"You're right there. In the past some companies scanned retinas, but in the end it was banned, because it was too intrusive a system, it violated a person's privacy. In this case it's you who decides to put your hand in if you want to enter. With eye scans, the only way to stop yourself being identified everywhere would be to shut them, and that isn't very practical ..." He closes his eyes and clumsily holds his arms out in front of him, a position that's even funnier because he didn't intend it to be.

Celia laughs, finally some humor, and Leo responds with an odd expression, surprised by his own actions. Sucked in by the holographic advertising that follows them everywhere, Lu has missed what happened and her daughter's laughter is what brings her back, though she's not quite sure to where. She doesn't know if she should scold her and, confused, she chooses to lose herself once again in the accelerated transformation of a fat woman into a model, thanks to the sculpturing massages that her top-of-the-line CraftER robot performs on her while she's sleeping.

"Personalized advertising"—Leo doesn't want to leave her out of the conversation—"it's one of the advantages of identifying guests. If you're interested in any of the accessories, let me know and I'll see if we can include it in the compensation you'll receive for your services."

Almost without realizing, they've stepped onto a mobile platform that slides along a web of ribbons that remind Celia of old road junctions, but on a smaller scale. Their destination must be preprogrammed, as the turns are made at the crossings without Leo having to do anything, until they stop on a very high floor and a pleasant female voice welcomes them to the recording room.

The visitors are taken aback by how large the space is, full of machines with not a living soul present. Or maybe there is someone here, on the far side of the room it looks like something's moving. It's a robot, which had been hidden among the instruments, and it's coming toward them.

"ROBco will help us record the signals."

"*Information:* Dr. Sus Cal'Vin has told me that you must contact her before we start."

When she hears Xis' mom's name, Celia's heart leaps and she looks around as if the others should have recognized the name too. But no. Leo only instructs the robot to make them comfortable in the recording booth, as he'll be right back.

She feels an instinctive rejection of the intruder who has whisked Leo away from her. Since she found out about it at school, this coincidence has bothered her. Things would only get worse if she were to feel the robot's metallic hands on her, rather than those of the young man. She would even prefer those of a disheveled old genius.

ROBco adjusts the recording chair and, with exquisite precision, places a helmet covered with spikes on her head, clearly demonstrating that it's been trained to avoid all physical contact. Celia had been so worried about all this a moment ago and in the end she doesn't even notice it, she's so focused on covertly watching the engineer, who's talking pretty fiercely, as if he were defending himself. The distance prevents her from hearing what they're saying and from clearly outlining the red patch on the monitor, but she can sense that there's a problem and she could swear it has something to do with her. While the robot presses buttons and pulls out antennas, she wonders how to create a good opportunity to ask about it.

As soon as he enters the booth, Leo notifies them, in a falsely routine manner, that Dr. Cal'Vin will supervise the whole session remotely, in accordance with the wishes of the president of the company, and will keep a copy of the parts that she feels it would be convenient to show him. He imagines they don't see that as any kind of inconvenience.

"Why would we? Isn't it normal practice?" says Lu, taking an interest.

"In the case of a minor, yes; it's necessary to provide proof that their rights haven't been violated. As you can see, CraftER always goes about its work with total guarantees."

Lots of guarantees, but she'd never felt she was in such a bad situation, thinks Celia, at the mercy of strangers. All this stuff about being watched over all the time for her protection is becoming insufferable, and it's really annoying when they use her age as an excuse. Who do they have to protect her from? Leo? She thought she'd practically be alone with him, subject only to Lu's inattentive gaze. And now she's found out she'll have to suffer not only the implacable company of the robot, but also the general surveillance of Xis' mom. She could at least have deigned to show her face.

Without thinking, she's allocated a face to the woman, complete with sunglasses, those unpleasant ones that don't let you see the person's eyes. Like Mrs. Pia's. She had never known if the cleaning lady was scolding her because she'd dirtied the stairs just after she'd mopped or if she was just telling her to be careful not to slip up. But that image doesn't quite fit in today's world. Suddenly she realizes she hasn't seen anyone wearing glasses, dark or not. It's hard to believe it's taken her so long to notice.

Leo appearing by her side cuts this worrying short, and almost cuts off her breathing too when she sees his hands coming toward her face ... but they go right past and land softly on the helmet she's wearing. Her shock must have been obvious, as he's quick to make it clear that the helmet must be well fastened in order to be able to adjust the sensors to the distinctive features of her brain structure. That's the first thing they'll do: run a series of tests so the nanosensors can stabilize themselves. As the process will last a few minutes and is undetectable to the user, who doesn't feel or notice anything, Leo uses the time to explain the tests they'll do next.

Celia glances over at Lu, expecting her agreement, but finds her more interested in the robot working with the instruments than in anything else. She'd like—but at the same time wouldn't like—her to pay more attention. Her mother would certainly never have taken her eyes off her, even for an instant, but right now she almost prefers to be left to her own devices.

Leo has sat down next to her and has started giving her instructions. At the same time as the stimuli appear on the screen, she has to give as many answers as she can. If there are unfinished drawings, she has to complete them in as many ways as possible; if objects appear, whether they're familiar to her or not, she has to imagine all the different ways they can be used; if there are situations of conflict, she has to give the maximum number of solutions. The young man's hands are still, but his eyes are widening just like Celia's. She's giving him all the attention she's capable of, and a little more.

Once the preliminary test is completed, the stimuli start to appear. The first one is a rod with a ball on the end. She has no idea what it's for, but it reminds Celia of her grandmother's cane, and then a pin with a giant head and, upside down, a hammer, a golf club, and the stick things jugglers throw up in the air, one of those pipettes they use in laboratories, a spinning top, a circus performer's torch...

"Should I continue?"

"Yes, yes, until you can't think of any more uses. I don't know what half the things you're saying are, but you can tell me later. The most important thing is that the signals are saved while you're thinking."

Having him so close to her is distracting, and she has to make a huge effort to concentrate on the stimuli and not stop answering. Only very occasionally does she allow herself to look away from the screen and charge her batteries in the sparks she can discern in the young man's eyes. During one of these brief diversions, right after having seen two lovers' hands intertwined in four clumsily drawn lines, she can't avoid looking down at his inert hand on the armrest and imagining her own flitting over there to interlace with his. She can almost feel the warmth, his fingers lightly brushing against hers, and she's dying to bring it to her lips and kiss it. The tension she feels around her mouth makes her shiver, and suddenly she remembers that Xis' mother must be watching them. Confused and suffocated she turns back to the screen praying that Leo hasn't noticed anything.

She keeps going as best she can, wondering who she can confide in. This is too much for her to deal with on her own, but Lu, despite being present, won't have noticed anything, and her mother won't be able to answer her. She doesn't really want to tell Xis either, now that her mother is mixed up in all this; maybe she'll just tell her about that part of the day.

She'd like to tell Silvana about it, but she's come here behind her back and she doesn't want to disappoint her by admitting that she ignored her advice. What a mess.

Seeing that her performance is decreasing, Leo points out that, if she's tired, they can finish another day. That's the best thing she could possibly have heard: the chance to come back, so, excitedly, she accepts right away.

On the way back to the aero'car, Celia dares to ask where his laboratory, or office, or whatever is, and the young man, pointing up, replies that it's actually in this wing of the building, on the eighth floor; if it weren't a private area, he could show it to them right now by going up that ramp. Being so close and then not being able to go there awakens a stronger interest in her, and she allows herself to hope that maybe, if she continues to collaborate, one day they'll let her go in there.

IV

THE UNEXPECTED EVENT

20

She worked so hard to convince Lu to let them do a session at the ComU, and now that the girl's finally here it seems to Silvana she's not all that excited about the visit. She didn't expect this. They'd talked about it so much and every time she was rewarded with a pair of twinkling eyes that communicated an enthusiasm for a dream they thought was practically impossible. And now that the dream has come true, all she sees are distracted, distant eyes, an empty gaze. Even turning around to look at Silvana seems to be a great effort. She's not her usual self.

She must not let the coldness affect her, as she well knows the only way to turn the situation around is to keep her own spirits up. The girl's visit was an important milestone for her too, and she's not ready to let it go so easily. She'd thought that just by walking around the communal facilities—the corridors full of people, many of whom stop to say hello; the library, with shelves and paper books; the arboretum and garden, the pride of naturalists and hikers—Celia's previous life would be evoked for her, and some feeling that may not have reemerged yet after her long hibernation would virulently flower. But it hasn't worked out like that. The girl is more absent than ever. Maybe it's all just as alien to her as the school full of mannequins or the inhospitable yellow of a house you can only reach by aero'car. What to Silvana is an abysmal difference may be no more than a nuance for an unfrozen person.

It occurs to her that, before taking the girl to the office, they could drop by the Emotional Stimulation Zone. That would make an impact, forcing Celia to focus on her surroundings, especially if there's a session running. With a little luck she might even tempt her to have a massage; that way she'll be able to read in the girl's body what she can't figure out from her behavior.

They pass through the middle S of the ESZ's holographic wall, as she always does, and sneak straight to the opaque zone so as not to disturb anyone.

"From here we can observe and hear them without them seeing us. The gas around us only lets the waves pass through in one direction."

She doesn't know if it's the surroundings or the technical explanation that has pulled Celia out of her haze, but for the first time those eyes look at her without the distracted veil that has been present until now.

"What are they doing?"

"Do you remember I told you I was an emotional masseuse? Well this is where I do group sessions. That's one of my colleagues. They've just started, do you want to join in?"

"Go out there ... and leave this great spot?"

"No, dear"—the word vibrates on her lips after she's said it—"I meant we could follow their instruction, but without moving away from where we are now."

There's a slight reticence in Celia's eyes, and Silvana knows there's a reason for it, though the girl doesn't. She won't take any risks, that's what she promises herself to help her calm down: she'll only read the girl's emotions, without stimulating anything.

"Will I have to lie down like them while you give me a massage?"

"Only if you want to, of course. It'll be a new experience, and you like those, don't you?"

"Yeah ..." She's still not sure, inspecting the floor where she'd have to lie down.

"It's pretty hard, isn't it? It's like that on purpose so you can feel every part of your body. We have to fight back against so much ergonomically correct furniture, all this anesthetic that's around nowadays ..."

She stops speaking mid-sentence when she realizes she's started off on a lecture that's totally out of place. She's anxious. She'd imagined this would be incredibly difficult to achieve and now it's within touching distance. Lying on the floor, barefoot like everyone else, Celia offers up her left foot just like she sees the others doing. Although Silvana's massaged countless feet over the years, today her hands are shaking like a beginner's. It's normal, she tells herself, she's never been faced with anything so small and tender. The soft, warm skin is completely different than the frozen, rigid soles Silvana is used to. Her pulse is beating strongly, all the way to the ends of her fingers as she places them on either side of the foot, ready to explore. Her extensive experience won't be much help to her now: this is virgin territory and she's not familiar with all its idiosyncrasies. Perhaps she won't even be able to interpret what she reads in it.

Her panic, however, dissipates as her thumb advances: although it's blurred, she recognizes the same sensorial structure as ever. It's possible the parts linked to sight and hearing are a little smaller while smell, taste and, above all, touch are more extensive and intricate, but she's not going to get caught up in this part right now, she wants to go straight to the deeper layers where the feelings live.

She doesn't have time to get there because Celia, who's been totally still and a bit tense the whole time, starts to wiggle around from left to right, until she breaks into unstoppable laughter. Silvana is so surprised she loses her train of thought, and almost loses her calm, since she's so worried she's hurt her in some way and is afraid they'll become the center of attention. But no, she immediately confirms that the opaque zone has protected them flawlessly.

"Sorry." The little girl has propped herself up on her elbows and her face is apologetic, with the added pathos of not being able to erase all traces of laughter. "It tickles."

The word hits her like a bullet. Now she really doesn't know where she is. If there's one taboo at the ComU it's mixing sex and massage, intimacy and professionalism, an especially critical barrier for Silvana, who's always scrupulously monitored herself in this respect. Today she's been taken completely by surprise: she'd never even considered this eventuality. Calm down, Silvana, this can't possibly be happening: she's just a little girl and you only touched her foot. Why is she talking about tickling? Maybe, without meaning to, she's stimulated some unknown area?

"Don't be frightened, sweetheart." She invites her to lie down and relax again with soft gestures. "Have you laughed like that before?"

"Of course! I had tickling wars with my dad all the time."

"With your dad?" She must have touched on some archaic practice. The sensory path is clear, now she needs to characterize the associated emotion. "Did you do it with anybody else?"

"No."

"And you liked it?"

She takes a little while to answer, observing Silvana very seriously as if she is trying to evaluate exactly how specific her answer has to be.

"I liked to start the war, I provoked it myself... even though in the end the sensation became a bit, I don't know, irritating. I really wanted it to stop, but I didn't cry or get serious, I couldn't stop laughing, I couldn't help it. Now I think about it, it's weird."

Silvana is getting excited about the mine of emotions she has before her, this capacity to surprise oneself, and the girl's attempts to explain every nuance leave her dumbfounded. She'll do whatever it takes to get to the treasure the girl carries in her mind.

"I promise I'll stop if it starts bothering you." Once again she's kneeling down at the girl's feet and is very carefully holding up the left one. "Do you mind if we continue the massage?"

"Are you only going to touch my foot? Don't people usually massage the back and the neck ...?"

"Sometimes, yes. But that's not my specialty." The eminently sensorial zones don't interest her at all right now. "If you want I'll do them before we finish." That's the least she can offer.

For a while now, isolated in their bubble, they've ceased to be aware of the movements of the other people in the room, and now Silvana, without a second thought, skips the initial exploration connected to the senses and concentrates instead on the small recesses between tendons, which have a direct connection to the limbic system. It's hard for her to decipher what she's detecting, because a strong signal smothers everything. She'll have to inhibit it in order to go any further, but the promise she's just made stops her in her tracks. All the more so when she looks up to see the girl's trusting face, giving herself over completely to the impulses of her hands. A shot of responsibility painfully oppresses her thumbs. She has the richest material imaginable at her mercy, but it's also the most malleable; any action would leave a mark and she doesn't want to hurt Celia in any way. But the reward is so large that it makes the risk seem small and Silvana's will is shaken. Maybe if she analyses that smothering signal she'll be able to find its origin, and then she could either eliminate it or take verbal action on the girl, avoiding contact stimulation.

The powerful signal has ramifications on most of Celia's organs and is conditioning the activity of her nerve tissue over a large radius. The impulses reverberate and echo in cycles. Silvana would classify it as an obsession if it weren't for the atypical location and the lack of an anxious component. Everything points toward a recent experience, something positive, that has caused a huge impact. She wonders if it might have something to do with her; this would have made her happy, but no, just last week Lu canceled her session.

"What goes through your mind when I press here?" She knows that the likelihood of her confessing it is small, because there's a voluntary blockage of the signal's escape routes, but she won't lose anything by trying.

"I hardly notice it. The truth is I'm a bit bored. You'll have to explain what all this is for."

Hurt by the comment and afraid that the girl will end up quitting their sessions, Silvana gives up on her ambition and opts for giving her a massage wherever she wants. With exquisite delicacy she explores Celia's back and puts her heart and soul into triggering highly pleasing sensations. It's been years since she did this, but she still recalls the technique and can remember which manipulations are most effective. Her hands slide everywhere, with special insistence on the base of the neck, happy to find such grateful skin, which bristles at the slightest contact and in no time at all takes on the perfect warmth. The pleasure comes and goes in a growing loop that reconciles them with one another, and even with the disparate worlds in which they must live.

When they come back down to earth, however, after spinning round and round in a shared spiral, Celia once again takes up that distracted stance, a little happier, yes, but just as incommunicative as before, and Silvana mentally prepares what she's going to tell Lu when she comes to pick the girl up. No, not now, Celia will be offended if she asks to speak to Lu in private. She'll call her, that's what she'll do, and she'll ask if they did anything out of the ordinary last week. No, not that either. She must avoid being so direct: she'll just stick to confiding to her that she's worried, she's seen that Celia's mind is monopolized by something that is preventing her from being her normal self. Surely she's noticed that for the last few days her daughter's had her head in the clouds. Maybe it has something to do with that bioengineer who was investigating creativity? With a bit of luck she may be able to scare Lu enough that she keeps him away from Celia, and she will have killed two birds with one stone. The second one is the most important, of course, canceling that smothering signal and obtaining free passage to Celia's deepest feelings.

21

Leo's on his way to his interview with the Doctor and has no idea what to expect. If this is all about comparing his prototype with those of his competitors he wouldn't have made him leave ROBco in the cubicle. How could Dr. Craft possibly evaluate the prosthesis without having it in front of him? And he's been so demanding: he wants the meeting to be free of recording and listening devices, like the one on the day of the convention, but this time in his quarters.

His memory of that first conversation is vivid—none of the following visits have been on that level—and he secretly hopes it will be repeated today. That day he proposed that Leo participate in his creativity project; who knows, maybe today he'll ask him to continue his work on wireless transmutation. He's kidding himself. The other day the Doctor disappeared without even saying goodbye, fed up with his verbal diarrhea. He can imagine what kind of impression the Doctor has of him now.

He sees a figure leaning against the wall at the top of the ramp his mobile platform is heading toward, and thinks it might be Dr. Cal'Vin. Just then a thought hits him: she must have sent the recordings of Celia to the Doctor, that'll be why he's been called over. He holds his breath and only when he's just a few feet away from the blonde woman does he breathe out again, relieved. It's not her. It must be a new employee, he's never seen her before. If there's anyone he doesn't want to bump into right now it's that idiotic neuropsychologist. She puts him on edge. Now he's even more pleased the meeting will be private.

One more floor and he'll be at the top of the tower. Now he needs to prepare his argument that the girl is essential to the project. The Doctor doesn't lack perspicacity, and if he constructs a decent argument, he's sure he can convince him of her worth. That is, of course, if he doesn't lose his head like the other day and manages to make himself heard.

Alpha+ greets Leo at the door and shows him into the dueling room, while at the same time asking him to please excuse the president for a moment as he's occupied with another visit. In case Leo wasn't surprised enough that they'd asked him to wait in such a private space, he is invited to sit down in the armchair in front of the table that had fascinated him so much the first day, and even more so after the Doctor demonstrated how the timeout button worked. He ecstatically contemplates the immense horizontal screen surrounded by swords and sabers of inlaid pearl that change color as you move around, and he sits down, his right leg next to the iridescent pink button sticking out of the front panel.

He can't get his head around why he's been left alone with such a valued artifact and he glances around the room, afraid someone might be spying on him. He'd give anything to be able to take it apart and inspect its inner workings. Maybe that way he could discover the secret of timeout and spare himself that awful feeling of having rented his brain out every time he comes in and out of the cubicle. He rests his arms on the armchair in an effort to control his hands. Given half a chance they would fly right over to the screen that seems to beckon them. Three human icons are blinking at the top: one without eyes, one without ears and one with all its senses, but cut up into bits, like a badly finished puzzle. Leo's eyes avidly focus on the textbox:

> There are now three monks living in the cloistered monastery, one of whom is blind and another deaf. A message informs them of an epidemic that manifests itself in the apparition of a red mark on the forehead, while at the same time it comforts them by saying that God wants to ensure the future of the order by having only one monk contract the disease.

Leo stops for a moment to take a breath. The Doctor lives in such an exotic universe. It's hard enough for him to understand the words "monk" and "God," and he has no idea what "cloistered" and "order" mean. He couldn't possibly compete in this game. The world of chess is more fixed and conquerable, not as absorbing perhaps, but the rules are clear and they're the same for everyone. Who makes them up here? Each competitor? No wonder he said it was a battle for only the best swordsmen.

When he leans over to keep reading, his hands wander to the edge of the table and his fingers rub up against a rounded, well-polished border. He likes this contrast of old and new: fleeting images trapped in a timeless frame, and never has that description been more apt…A unique setting to match what appears on the screen:

> In order to accomplish the heavenly dictate, for the time being the superior lifts the ban on communication, but they can only speak under strict regulations: they will stand in a line, and in order they will be allowed to say "diseased," "well," or "I don't know." Only once. Then the monks will go straight back to their cells and those who are not convinced they are well will commit suicide to avoid infecting the others, thereby saving the community.

The last word, which has an anti-techno flavor, makes him suddenly lean back as if he might catch something. Feeling cowardly, he again glances from side to side wondering whether they're testing him, but he can't see any cameras or microphones around. What does it matter if they spy on him, right now he wants to finish reading:

> As usual, there are no mirrors, lakes or any reflective surfaces at the monastery where the monks could see their reflections. Can the three monks guarantee the future of the order by placing themselves in the right order in the line? Three points for a correct answer within a five-minute period.

Now that he's finished reading, it doesn't seem that difficult. Let's see. There are seven possible combinations of sick and healthy monks. If they all decided to look after themselves and they didn't commit suicide, they would have only a one-in-seven chance of surviving. One must trust in the intelligence and desire of everyone to want to save the order, and know how to exploit each person's capacities. They have two sources of information available to them: the marks and the answers of their colleagues. The deaf monk, whatever position he's in along the line, will only know that he's well if he sees two red marks. The one who can see and hear is the one who receives the most information, and also the one who can give the most. The blind one doesn't add anything, he might as well be mute. That's it: he's found the way to save the order and, without thinking, drags the icons into their correct position on the board.

"Correct response, but outside the time limit." The reply fills the box for a second before making way for the next question: "Which monk would you rather be?"

Leo automatically presses the icon of the blind man, the only one who would be saved every time as long as he's not unwell. Such irony: the one who has the most disadvantages and provides the least information is the one who benefits the most.

Leo is so absorbed in answering questions that he'd never have guessed he's been waiting almost two hours when the Doctor's deep voice booms out of a far corner of the table and makes him shake from head to toe:

"Damn kid, that's enough. Who gave you permission to touch anything?"

Leo stands up so quickly that he hits the pink button with his knee and bounces back into the armchair. On the second attempt he manages to stand up and babble an apology. Into thin air, of course, since there's no one in the room.

He stands there for a second, disoriented, not daring to sit back down and regretfully looking at the table, until the Doctor comes in and makes him jump once again:

"You're a bold one, alright, and quicker than the other two."

Seeing him disheveled, wearing a black bathrobe and slippers, Leo wouldn't have recognized the Doctor if it weren't for his eyebrows, as vigorous and emphatic as ever.

"I came as soon as you called for me, yeah. Who are the others?"

"The deaf one and the mixed-up one, of course. I hope you're not offended about me blinding you."

"Blinding me?"

The Doctor takes a quick look at the table.

"Did you press the button as well? You are a quick-fingered one…" He comes closer and presses the button, inviting Leo to sit down.

"I knocked into it when I stood up." He apologizes while he settles himself in the chair facing the Doctor. The monk problem resurfaces in his mind making it clear that there must be some deeper meaning he's missing. "I guess you called me over to evaluate the prosthesis."

"Exactly. And you'll be pleased to hear that you've come out on top in the test, just like the blind man. Now it all depends on whether or not you have the red mark."

Leo feels like he's been challenged to a duel that is going straight over his head. He doesn't know what move he has to make, what he has to say.

"But you haven't even seen the prototype."

"So gifted in some areas but so short-sighted in others. What would I get out of examining that heap of bytes? What I'm interested in is its effect on PROPs, on you in this case, in order to extrapolate how it would affect me if I implanted it in Alpha+."

"Then you need to see me with ROBco, don't you?"

"What do you think I waste my time watching over you in your cubicle for? To spy on what you're doing? I don't have any desire to be a policeman, you know, or a voyeur. I don't give a shit about your life, what I wanted was to see you interact with the robot. And I've seen that already, but I need to check the permanent effects, what's left when he's not there. It's not about making us dependent on lumps of metal."

It's strange to hear the president of CraftER say that, Leo thinks, even if it is in private with no one listening in. Eager to continue hearing the Doctor speak in this direct way, without the metaphors he used before, he stares intently into his eyes in order to flatter him with his attention. And to spur him on that little bit more, he stresses the point:

"Of course not, the most important thing is ..."

"To design stimuli so that they shape us the way we want, the way I want, dammit. And then do without them. Your prototype is heading in the right direction, but it still has serious drawbacks. Today Alpha+ will send over the list of features that are missing. Alph," he shouts.

A sudden flash pops into Leo's mind. He's just another stimulus for the Doctor, and when he's fixed the prosthesis's deficiencies, the Doctor will do without him, just like he said. He decides to take a risk:

"This table is the best invention I've ever seen," he says, bringing his hands closer without daring to touch it. "Will you get rid of it too someday?"

"You are incisive and obstinate boy, aren't you? I'm glad you know how to appreciate excellence and ambition. That's why I picked you. The timeout device is the best I've ever come up with ... that anyone's ever come up with! But…not only would the time need to have run out, I would also have to be dead before I agreed to get rid of this table and let you enjoy it. Only the table, of course, since even with me dead you wouldn't be able to use the timeout device," he adds with an enigmatic smile. "Ah, Alph, you're here already. From now on you will give maximum priority to what Mar'10 asks for, starting with the list you have to send him. And you, get a move on, I want that prosthesis ready next month."

Leo leaves the room feeling proud that the Doctor recognized his talent, but at the same time worried about not seeing in this recognition the reward he hoped for: a guaranteed future.

22

When she gets up Celia can sense that something's not right. Everything looks the same as any other day: her dress and calendar are there next to the bed, as if she were going to school, and she can only hear ROBbie doing his chores around the house. Could it be that ROBul's forgotten to wake Lu up? She's about to go out into the hall when she stops herself, remembering she's naked. What a nuisance having to sleep with no pajamas on, she'll never get used to it. She covers herself with her robe, despite not having bathed yet, and heads straight for the machine room where, indeed, she finds ROBul connected to the DOMOsys carrying out the daily maintenance tasks.

"Hello. We're going to CraftER today, don't you remember?"

"*Information*: The interview has been called off."

"What?" Her voice comes out tiny and frail like the world has just collapsed on top of her. "Why didn't anyone tell me?"

"*Justification*: Lu didn't instruct me to tell you."

"But ROBbie knows about it, right?"

"*Affirmative*: I have to inform him of all the schedule changes that affect him."

"What day has it been changed to then?"

"*Information*: none. Lu said she was no longer interested in the project."

"Lu? It was her who canceled it?"

"*Affirmative.*"

She's enraged that Lu could have done such a thing without consulting her first, that she cares so little about what she might think or feel, but at the same time a small relief creeps up inside her: at least it wasn't Leo who refused to see her again.

While she's getting dressed, having breakfast, traveling to school, sitting before EDUsys, standing in front of the wooden figures in the socialization room, even when talking to Xis, she doesn't stop thinking about it. She's already lost her parents and she's not ready to let go of the one person she's ever excited about seeing. Well, maybe he's not the only one, there's Silvana too, but it's different with her, she'll always be there ... The idea comes as a revelation while her friend is going on and on, complaining about never being allowed to go into CraftER: they'll go there! The two of them, on foot. The company is only two blocks away and the exit in the school bathroom is never monitored. They often stay there chatting during the topic extension session. Not even their ROBs will miss them during that short period of time.

"And what about the entrance code?" Xis objects, all worked up.

In light of the prospect that has just opened up before her, there is no obstacle that could possibly stop Celia. She can take advantage of the fact that they registered her hand the other day and that their identity has to be protected since they're minors. The voice at the entrance made it very clear. They can't be identified, and, therefore, they're indistinguishable. They can both go through as if they were one person. And, if it doesn't work, they'll get back to school earlier. It's that simple. They've got nothing to lose.

"And what if we get caught?" her friend insists, obviously scared. "And what if we get lost?"

"I would have thought you'd want to see your mom at work ... But, if you're too scared, we can just forget about it."

Celia's not at all convinced she can forget about it. She's not convinced she could escape on her own either ... and even less so now that Xis knows about it. Why did she have to tell her about it so soon? She could have thought it over a bit. But what does it matter anyway, she would have had to tell her at some point if she didn't want Xis to raise the alarm when she couldn't find Celia anywhere. In fact, Xis would have to hide while she was out, because if they saw her wondering around on her own they would suspect something was up and would send the SEEKer to find her.

The advantage of sneaking out on her own is that the code would definitely work. But…how would she explain to Xis her reasons for wanting to go there? With the altruistic motive of visiting Xis' mom out the window, her feelings would be revealed, and just imagining them being exposed like that makes her feel awful. Without a doubt the best option is for both of them to go. So Celia keeps on trying, determined to patiently defuse all the but's that, one after the other, her friend keeps coming up with.

When she's succeeded in convincing her and they're in the bathroom ready to go, Xis offers up one final bit of nonsense:

"Do we have to go right now? Ok. I'll just tell ROBix we're leaving and then we can go."

"No, I told you already"—many more moments like this and she'll regret having persuaded her—"our ROBs can't know anything about this, they might stop us or tell on us."

"My ROBix…never!"

"How do you know? Have you ever tried it? Anyway, we already agreed, we're not going to tell them anything."

"You're right, but…"—she's all stressed out—"it always has to know where I am."

"Why?"

"Because…I don't know, how else will it keep an eye on me?"

"Xis, if they're watching us they'll never let us go to CraftER."

"Yeah, I get it, I get it…but I've never done this before. How will we find the way?"

"Don't worry. You want to see your mom, don't you?"

"Yes…"—a pair of hopeful eyes eclipse any last minute doubts.

"Well, it's really close." Taking her by the arm, she leads her to the exit without any resistance.

Outside there's not a soul in sight. The only movement they can detect is going on above their heads, where a swarm of aero'cars crosses paths time and again, creating a massive spider web that, although fleeting, is, in Celia's eyes, clearly drawn. She's never seen them from this far down and, for an instant, the perfect order of their trajectories entrances her. It seemed so chaotic from up there.

Mechanically she's taken a few steps forward, walking along close to the wall as if there were still sidewalks to keep to, when she realizes Xis isn't following. When she sees her standing still in the middle of the road gazing lifelessly ahead, she turns back to take her arm again and tells her very quietly that there's no need to worry, they can see CraftER from here: it's that golden building, wide at the bottom and narrow at the top, that sticks out between two dark tower blocks. The pinecone, as she'd thought of it the other day. When they get closer they'll be able to see the lattice of little cells where the aero'cars dock.

The pavement, which is covered in bits of glass and scrap mental, perhaps left over from accidents, contrasts with the lustrous constructions that rise up on either side of it. The majority don't have entrances at street level, and others, the most dilapidated ones, have bricked-up doors. For a moment, the suspicion that they may not be able to get into CraftER from the ground level worries her, but remembering the other day's visit restores her confidence; she's one hundred percent sure that the aero'car flew down to the ground and that Leo greeted them there, on the ground floor. It must be because the most powerful companies have made the street into their own private parking lot whereas the others have had to move theirs upstairs.

All of a sudden, an aero'car flying lower than normal makes her leap backward, letting go of Xis' arm, who keeps walking unperturbed. The pilot slows down, perhaps due to Celia's brusque movement or perhaps because it is simply intrigued to see two people in the road. Celia is convinced they're observing them despite the fact that she can't see inside the vehicle. She hurries over to her friend to tell her about it and is surprised to find she's not aware of anything having happened.

"Which one? That one? It got close to us and you let go of me? Did you want me to get hurt or something?"

"Please, Xis, it was a reflex... of course I didn't want you to get hurt."

"But you saw it and you didn't tell me. ROBix would have warned me... instead of abandoning me." She starts whimpering.

"What are you talking about? Abandon you!" She doesn't understand why Xis' reaction is so over the top. "I let you go because it made me jump, that's all."

"That's horrible: you see danger coming and you leave me on my own. Why did you want me to come? So you could use me as a shield? Let's go back right now!" She's getting all worked up.

"Please, Xis, calm down." She tries to take her arm, but her friend pulls away. "If I hadn't told you just now you wouldn't have even known an aero'car got so close to us. I could get upset too and start shouting: It's impossible for you not to have seen it, you're lying!" This performance is actually pretty convincing. "But I accept that we're different, that it makes me jump but you don't even notice it. And who's saying it was dangerous? It wasn't dangerous at all. The best thing to have done was just ignore it and keep moving forward, like you did." At least she's managed to calm her down and make her listen.

"Now you're trying to flatter my circuits so we can go to CraftER, right? You always have to get your own way."

"Not at all. I really think that. Can't you see that I was born a century ago and I'm not made for life in today's world? It's logical that you would act more appropriately."

Convincing her was easier than she thought. Although Xis did spend the rest of the journey gripping her arm so hard it was almost numb.

Just as she imagined, the enormous parking lot is spread out like a rug before the endless facade, where, as they walk along, they find ever more openings on ground level. They're all identical. She'll have to give up on the idea of finding the one from the other day and risk trying her luck with a random one. They've already passed about a dozen when they decide to try one that's in an open area with no aero'cars parked nearby. That way there'll be less chance of being seen.

She takes advantage of explaining the strategy to Xis to liberate her captive arm and, a few minutes later, she's putting her still sore hand into the cavity next to the door. Today the voice doesn't take her by surprise: "Identified as Leo-1." It works!

They stand side by side and try to go through together, but an invisible barrier stops them in their tracks. It's not as firm as a wall, it's more like a membrane that, when it touches them, saps their energy. The panic brought on by feeling trapped dissipates when they step backward and, what a relief, they don't find anything stopping them. They try again in case their synchronization wasn't good enough, but the same membrane spits them out again, and this time a voice booms "you must go through one by one."

She tries to convince Xis that, to check if the code works, she'll go through once on her own. She'll just go in and come right back out again, she promises, Xis won't even notice she's gone. Celia's afraid her friend will go off on her like before and refuse to split up but, luckily, she is so keen to enter that she accepts the promise with no argument.

It feels strange to go through the opening so easily, but the shock comes when she tries to go back. She can't. The membrane from before has appeared behind her and is draining her energy. She tries to find a solution in what she can see around her, while at the same time avoiding looking at Xis, so as not to worry her. A corridor just like the one from the other day stretches out in front of her, and she can make out a series of openings on either side of it. She really is a fool. Of course each one must have a corresponding cavity inside the building and, yes, hers is right next to her. She puts her hand in and "Identified as Leo-1" rings out like a blessing before she heads back outside.

Encouraged by her success, she puts her hand back in and lets Xis through the opening. Both of them have a protected identity, the system should be incapable of telling their chips apart. And, precisely, it doesn't distinguish one from the other and her friend smiles at her from the other side. She's taken a risk: now she needs the system not to have saved any record of "her" just having come in and allow her to open the entrance again from outside. Just imagining her friend imprisoned and being unable to help her sends a shiver down her spine and makes her still numb hand smash into the side of the cavity. Now it does hurt, she's bleeding and everything, but the only thing she's worried about is whether the wound might distort the code. She dries it off the best she can to hide it and, hesitantly, puts her hand into the cavity. "Identified as Leo-1." Celestial music, once again.

She crosses without any difficulty and is met with Xis' serene expression, unaware of the risk she's just taken. She's certainly not going to tell her about it. The corridor is as silent as ever, and they move along pressed up against the wall so they can take refuge in one of the openings if someone comes along. Celia wants to get to the ramps from the other day and go up to the eighth floor, Leo's. It's easy to convince Xis that, although she's not entirely sure where her mother is, it will be easier to spot her from high up, and at the same time they will be less exposed.

They must have come at just the right time of day, because they don't see anyone the whole way there. They're up at the top and the view hasn't changed much: on the side they're on, there's the same series of openings as ever, which, as they've discovered, open onto the platforms where the aero'cars are kept, like pine nuts inside the giant pinecone; and, on the other side of the central gallery, a line of doors. She'd give anything to know which one was Leo's. They crouch down next to an opening, looking out for any kind of movement and, surprisingly, it seems like everyone's been waiting for them to arrive in order to start the show: the door farthest from them opens up as if by magic without letting anyone through, while from the opposite side of the space a great clamor of voices and footsteps announces the arrival of a whole army of people. They both try to hide at exactly the same time, but they can't find anywhere to go, and Celia has the brilliant idea of going through the opening and watching the show from a safe hiding place. She's already had plenty of practice putting her hand in them as many times as necessary.

These portholes that allow you to spy on everything with a low risk of being discovered are fantastic. They remind her of her mother's ring, and squeezing it in her pocket makes her feel safer. Behind her, one sole aero'car nestled into the platform has turned the space into a closed, almost welcoming, area. They've positioned themselves on either side of the aperture and stretch their necks out toward each other in order to see out of it. Just as they suspected, a whole load of people are filing past them, and, suddenly, Xis starts jiggling around uncontrollably and softly squealing: "my mom," "my mom." Xis' mother isn't exactly how she'd imagined, but it isn't much of a stretch to put her in those dark glasses from the other day. Although she's smiling at her conversational partner, she still comes across to Celia as unfriendly, and gives her bad vibes. As usual, she's getting in the way and distracting her from her search for Leo. She returns her gaze to the door they're coming in through and is frozen in place by what she sees. Over there, it's definitely him, he's not talking to anyone and he's looking down at the floor. She longs to go out there and make him look up, talk to him, ask him to show her where he works, but she can't…She would like to at least be able to hold on to Xis, show him to her and tell her it's him, but she can't do that either. She feels like an idiot just standing here, useless, not daring to do one thing or the other.

She doesn't have to feel like that for long because right away alarm bells start ringing inside her head: a man and woman are coming toward them. With some difficulty she manages to push Xis down on the ground and get them both to huddle up behind the aero'car, when the metallic voice starts booming and the two strangers cross the threshold. Without exchanging a word, the woman gets into the device and it starts up. The shock is tremendous. For no reason at all Celia had supposed the man would pilot the machine, but now she can see that's not right, and she is hit by a worrying suspicion: maybe the pilot is a ROB and it's been observing them the whole time. The man has gotten in too and she's still wondering how she could possibly have neglected to think of this when a powerful shaking alerts her to their immediate danger. Frightened, she grabs Xis and drags her over to take shelter by the opening, just in time to see the huge machine fly away, and an enormous hole, which this time is not solid at all, opens up before their eyes.

The spectacle is so magnificent and unusual that the girls feel drawn to the edge of the platform in order to take it all in better. The lattice of tensors and moorings that had been holding the aero'car in place serves as a railing for this improvised balcony. They watch the ship get farther and farther away, majestic, until it is lost among all the others, swallowed up in that highly organized collective movement. When they look down it's hard for their eyes to adjust to the brightness of the facade, but little by little the different levels of platforms, with or without vehicles, start to emerge in a kind of giant screen of light and shadow. Celia is entranced, but her friend's cries bring her back in an instant:

"I'm going to fall, I'm going to fall!" Xis' hands grip the pole that separates them from the abyss, turning her knuckles white.

"Calm down, Xis, don't look." She gently takes hold of her from behind and makes her turn around. "It's not that different than what you see every day from the aero'bus that takes us to school."

"But there's no protection here! I can see there's nothing, nothing"—she whimpers, terrified, stretching her arms out as if they too were part of the horror. "I want ROBix right now. I want to go back."

"Okay, Okay."

She pushes her friend toward the opening and, after checking there's no one on the other side, Celia puts her hand in the cavity. She's so anxious to hear that soothing voice that the machine's silence feels even longer than usual. And longer still. Shaking now, she puts her hand back in. Nothing. The system has stopped working, it's like it's disconnected, dead. She moves over to touch the opening and, when she comes up against that transparent membrane, more solid than ever, a gnawing feeling starts in her stomach. Could it be possible that it can only be activated when there's an aero'car parked here? She's really messed up this time.

She wants to hid it from Xis, make her believe that her mother will walk past again in a moment, that it'll be nice to wait for her, but her friend has already discovered the way is barred and turns to her with fear written all over her face:

"Open it, open it! I want to get out." She strikes the membrane with both hands, beside herself. "I want ROBix. It'll know what to do. I need to talk to it..."

"Calm down. We're not in any danger. We'll just wait here quietly until another aero'car parks and we'll get out. We could be out really soon."

"We'll never get out! You're from another century, you have no idea how anything works." She suddenly switches from rage to inconsolable sobbing. "How could I have let myself listen to you?"

Any desire to fight back disappears when she sees Xis curl up in a corner with her legs pulled up against her chest and her arms wrapped all the way around herself. So helpless that compassion more than anger is inspired in Celia.

"Stay calm," she whispers, huddling up next to her. "Nothing bad is going to happen. We can have a good time telling each other stories while we wait. I'd rather get out without them seeing us"—the words "Identified as Leo-1" are still ringing inside her head and she wouldn't want to cause the bioengineer any problems—"but if you don't feel well..."

Her only reply is two unseeing eyes. Celia won't get anything out of talking to her, and she feels too tired to make more of an effort than that. It'll be better for both of them if she gets comfortable and tries to relax, as it could be a long wait.

23

The first person to notice the girls' disappearance is Lu, because when she goes to pick Celia up from school the SEEKer can't find her. Today, unusually, she's decided to go there herself. ROBul said she should, since the girl got so angry that morning when she found out they wouldn't be going to CraftER. As she's turned up without telling anyone and she's unaware of how the place normally is at this time, she's not immediately worried, and trusts that Celia will appear within a few minutes to get on the school aero'bus, like she does every day. When the bus comes, however, and the teacher confirms that Xis is missing, too, she can find no comforting thoughts in which to take refuge. Her daughter, missing! And now what? Which procedures will she have to follow? What will she tell her friends ... they'll think it was her fault; thank goodness it hasn't happened only to her and that famous doctor's daughter has disappeared too.

Lu listens as the teacher relays the situation to Sus Cal'Vin, who can't believe what the teacher is telling her. She stubbornly repeats that it's impossible and they should conduct a proper search. They've caught her when she's about to go into a meeting so she can't deal with the situation right now, and it's the school's problem anyway, not hers. Of course, she demands to be kept up to date at all times, and if they need any information about her daughter they can ask her friend Fi. This is the last thing Lu needed, putting their shared friend in the middle.

For now, Lu and the teacher alert the POLis and follow the early developments from the school. In the initial phase, the missing girls' images are disseminated to every optical sensor in a radius proportional to the time that's passed, in concentric waves that get progressively farther out. When any sign of detection is produced, the images are shown to the complainants so their validity can be confirmed or denied. One, two, three images: Xis in the middle of the road, Celia next to a wall, both of them walking arm in arm along the pavement. Confirmed, it's them. Together and alone. They left of their own volition then, so the school is not responsible.

The teacher washes his hands of the situation, and Lu continues to follow any leads from inside her aero'car, helped by ROBul and ROBbie, ready to go and pick the girls up from wherever is necessary. The epicenter of the search waves is now focused on their last known location, in the street near the school. Next image: Celia, scared, watching a ship come closer to them. What if they've been attacked and the next photo shows them on the ground, horribly injured, what will she do? ROBul, always attentive, calms her down: she won't have to do anything, the POLis will take responsibility for calling an aero'ambulance. But that won't be necessary, because just then they see them walking along arm in arm once again. They decide to follow their path step-by-step and, with each image, they are getting closer and closer to CraftER.

Stationary before the luminous facade, they're waiting for a new photo to show them where to go. In vain. The clues stop here, the POLis informs them, and when they ask for an explanation, they are told that it doesn't have access to any images taken inside businesses; in order to see them, they will have to request the corresponding permits. They can't believe they haven't realized until now, the girls must have gone into CraftER. It's totally logical, ROBul states as he initiates the authorization process: Celia woke up thinking she would go there, and she's gone! But Lu makes him shut up: ROB logic, no daughter would think to disobey her mother; someone must have forced her. Maybe that Leo. Otherwise, how could they have gotten inside?

She can't decide what the next step should be. She would prefer to see the recordings before intervening, but she doesn't know how long it might take. The courageous solution, now that she thinks of it, would be to contact Leo directly, her friends will reproach her for it if she doesn't; the problem is that she's not sure if she should ask him for help or threaten him. Who knows, he might have kidnapped them to suck out their brains. She'll really be putting her foot in it if she gets it wrong.

After all this she'll end up bald before her time. She's got to calm down, by any means necessary. Didn't her beautician ban her from getting stressed? So someone else should manage this. Just as Dr. Cal'Vin delegated to Fi, she will pass the responsibility on, too ... not to her friend, she doesn't trust her, but to a professional. ROBul will find one. These CraftER people are really mean-spirited not letting it in. She's sure it would sort this out.

He really is efficient, her ROB, he gives her some proposals right away. Let's see: the home tutor, of course, why didn't she think of that? She's a psychologist and must be used to resolving conflicts like this one. Besides, she caused the incident by suggesting they cancel the interview, it's only fair that she solve the problem, and sort out Celia's bad mood while she's at it. That's the last thing she needs, to have to fight with the girl, and put up with her unpredictable reactions. Of course if she has been kept there against her will, the tutor will take all the credit. Who cares. The most important thing is that her stress levels have plunged.

***

It's hard for Silvana to understand what Lu is asking for; with some difficulty she works out that she's lost the girl—it's Silvana's fault, of course—and she doesn't know if she has to go and negotiate with a kidnapper or convince a teenager who's sick of her mother to go home. Since it's Celia, however, she wouldn't miss this opportunity for the world, one in which she might have the full range of the girl's emotions at her fingertips.

The emergency code X327-Cel-Xis-345P, which the POLis has assigned to the case, gives her immediate access to an emergency aero'car, which, since it has priority, takes off immediately, flying toward the exit and out into the open. The sudden brightness forces her to close her eyes, while ROBul's disembodied voice carries on, picking apart the details of what has happened. It angers her to learn that, behind her back, they'd volunteered themselves for that pretentious bioengineer's project. That will be why Celia was so inscrutable the other day, she had to keep it a secret. And how uncomfortable for her, poor thing, to resist Silvana's massagist attacks, which had taken her to the limit of what she could control. She's ashamed that she's inflicted such a torture on her, but at the same time, deep down, she can feel a joyous warmth. She knows that, with the secret revealed, she'll be able to win back the girl's trust. Intact. Having the weight lifted off her shoulders automatically makes her open her eyes, but only for an instant, just long enough to be blinded.

Back in the dark she recovers her train of thought, and ROBul's words penetrate more deeply. Now she understands: the warning nudge she gave to Lu scared her to the point of making her cancel the interview. She didn't confess that they'd been to CraftER, however, despite being a seemingly transparent person. If at first she thought that intervening in this delicate situation might win her some kind of points in the girl's eyes, as she goes through the facts of the morning's upset and the images of the girls in the street, she builds up a scenario in which she is the intruder who's there to oppose Celia's wishes. She's terrified of the possibility of ending up on Lu's side.

More so than her eyes, it was her thought process that had been blinded before: what an error to think that with the secret out in the open any distance would be erased. What does it matter that she went to CraftER before? What really matters is what happened there, and the mark it made on Celia. And she's all too aware of what she read in the girl's brain to keep on denying it: a positive emotion tinged everything. Everything. She finds it difficult to infer any interpretation that could make her feel better. Despite the antagonism she feels toward CraftER and its lackeys, she struggles to blame them for the disappearance. As a worst-case scenario, the boy will have gotten something out of it, who knows how much. She doesn't even want to think about it.

When she opens her eyes again, she's in front of CraftER. She has to admit that, when it comes to organization, the pro-technos have had their successes. She'd heard of this exclusive ultrafast form of transport, and, now that she's had the chance to try it, she understands why they sometimes call the anti-technos parasites: because, although they criticize the pro's advances, they do end up taking advantage of them. She almost blushes at having had such an improper thought, and she gets out of the aero'car determined to stick to her guns.

Lu and the two ROBs are waiting for her with the connection to Leo ready. Since they've already told her everything, they don't need to waste another second and, once she's on board, they urge her to start the negotiation—in her name only, since Xis' mother got very anxious when she heard the girls might be inside CraftER, and ordered them in no uncertain terms not to do anything until they received authorization from her. They'd better make sure she doesn't find out they ignored her.

The communication request is accepted right away, but right at the beginning of the protocol process, they are denied access because the emitter can't be identified as the line owner. ROBul doesn't have time to fix this before the connection is reactivated and Leo appears on the monitor:

"Good afternoon, Lu. Have you changed your mind?"

Does he have no shame?, Silvana thinks. Whether the girl has gone because she wants to or because she's been forced, it's outrageous that he would rub her mother's face in the fact that her decisions count for so little. The delay in response makes the boy realize he's not speaking with whoever he thought he was, and his worried expression immediately becomes annoyed. He's recognized her. Before he cuts off the communication she intervenes:

"I represent Celia's mother." She says this with the solemnity of a lawyer. "The girl needs to come home right now. If you need more recordings, you can make a proposal and we will consider it."

"Tell her that I'm pleased she's changed her mind. It's very important research, not only for the company, but also ..."—he shakes his head—"if you don't believe in this, why are you acting as intermediary?"

Now she's the one confused. He's just accepted that mediation is necessary, but at the same time it seems too spontaneous to be a kidnapper following a premeditated strategy. Either he's a great liar or something doesn't add up. What does it matter what she may or may not think of the project?

"Of course I personally wouldn't lift a finger for this research. It's the girl I'm worried about."

"Me too, trust me. She has a unique mind, and I'll do whatever it takes to record it again."

Silvana feels a shiver run down her spine: she might have said the same thing, in other words, yes, but exactly the same thing. Like a pro-techno. She hates herself for thinking this and at the same time feels too close to her opponent to be able to carry on fighting. Like him, she wants access to the girl's feelings right now. She envies him, she really does envy him! The girl has chosen to be with him. Now she's sure of it. And in an outburst of generosity she makes an offer:

“Prove to us that the girl is fine and maybe Lu will feel a little better.”

“Sorry, but I don’t understand. What do you want me to show you? The recordings from the other day? I can assure you they don’t have any lasting consequences. How can I demonstrate that she’s fine?”

“Let us see the girl and talk to her.”

Leo takes far too long to reply, and when he does she can hardly understand him.

“You don’t know where she is…Why didn’t you say so from the start? We would have saved…That’s it, isn’t it?”

“We know she’s inside CraftER.”

“Do you have pictures?”

“We’re waiting for authorization.”

“I’ll check on it now. Do you have any idea who she came in with?”

“Really, she’s not with you?”

“No. Let me know if there’s any news. I’ll try to speed up the permissions process.” He abruptly cuts off the communication.

The exchange of feelings between the two women is cut short by an urgent warning from the POLis: deceased matter has been detected on a public highway, at point 7N28E, in front of the CraftER building, and the organic remains are in transit on their way to the Center for Recycling of Human Material. As soon as they’ve been analyzed they’ll be informed as to whether they correspond to the two missing persons in the case X327-Cel-Xis-345P.

No matter how many times she opens her mouth, Silvana can’t manage to articulate a single world, while Lu complies with ROBul’s request to notify Leo to see if they can get any more information.

24

It feels like she's only had her eyes closed for a second, but when she opens them Xis is no longer sitting beside her. She can't see her anywhere else on the platform either. It's hard to believe that an aero'car could have landed and taken her away so stealthily. It must be that someone, probably her mother, has managed to unlock the opening from inside. She stands up and goes over there expectantly, but the silence in the corridor seems more hostile than ever. Why has she been left here? No matter how angry Xis was, Celia can't imagine she would be capable of abandoning her like this. It's true that she's gotten her into a real mess and the poor girl was really anxious; so much so that in order to calm her down and make up for what she'd done, Celia had to agree that she would let the girl cut her hair off when they got out. Hair is one thing, but trapping her in this hole is something else entirely... Suddenly, Celia's face lights up: who's saying she's trapped?

She throws herself at the cavity and sticks her hand in, desperate to hear that restorative voice say "Identified as Leo-1." But nothing's happening, the system remains silent. Skeptical, she leaps at the opening, but her hands and forehead crash into the membrane that, due to the sudden momentum, has become hard and hurt her. She's not thinking anymore, just worriedly pushing on the door looking for a crack, a sign that indicates she'll be able to get in. Although she's not any more trapped than before, she feels like she is.

As a last resort, she grabs hold of her ring, takes it out of her pocket, and tries to calm herself down. Her mom will show her what to do. She just needs to find a place where she can point it at the sky. Trying to find a better view, she shakily moves over to the edge of the platform, where the unexpected contact with the pole that she'd used as a barrier brings back the image of a hysterical Xis shouting "I'm going to fall, I'm going to fall." A lump forms in her throat. For a moment, she's in her friend's skin and the vertigo makes her feel like she's about to tumble into the abyss.

She squeezes her eyes shut just as tightly as she's gripping the ring and carefully moves backward, finally sitting down against the wall. Her heart is beating wildly and her limbs have gone numb. She wishes she could stop thinking what she's thinking. Stop feeling what she's feeling. That she'd never thought of coming here, proposing it to… "No, Xis, no!" Her shout comes out already muffled. It's just not possible her friend could have become so desperate that she'd throw herself off the platform. There must be another explanation, but right now she's too depressed to see it. Her mother will help her. Together they'll build barriers to keep out the bad thoughts, like they did when she was sick, they'll get rid of this terrifying feeling of falling.

Looking through the hole in the ring, she crawls along by the wall looking for a piece of sky, even if it's just a tiny one. She has to lie down on the ground to find a glimmer of blue sky, and, as soon as she manages to line it up: Mom, Mom! A flood of tears once again blinds her. It's horrible, Mom, what if Xis is dead? It's all my fault, she didn't want to come, it was me who pushed her into it. Oh, no! What am I saying? Not pushed her, no, I never touched her. Oh, Mom, I don't know what I'm talking about, my head is spinning. And Leo? They'll fire him… that'll be my fault too. I've ruined everything. How could I have ever thought to come here? I've changed so much… I never ran away before, did I? "If you get lost, don't get all worked up, just think," that's what you told me when I was little and we went to crowded places. I have to think… but there's no one here, no guards to ask, no stores to go into. My only hope is for someone to walk down the corridor… and, from where I am, I won't even see them! She stretches herself right out so she can look through the opening and a woman's face disfigured by the membrane almost scares her to death. Reining in the instinct to hide, she recognizes Xis' mother gesturing at her. She's come back to save her. Her first thought was right after all, the easiest solution ... why did she have to imagine such a monstrous situation?

From the signals she's making with her hands Celia understands that she's telling her to wait a moment, she has to go and get something and she'll be right back. Celia gets the impression that she's also asking her to hide so she won't be seen. She must want to get them out of CraftER without anyone noticing. Celia bore such a grudge against her before, but now she'll even be thanking her: it would be great if Leo didn't find out about what's happened.

As if she's conjured him up, the young man's outline appears framed by a door right in front of the opening, and, even though she hurriedly pulls back, she figures he's seen her. Her desperate attempt to avoid detection by flattening herself against the wall turns out to be totally pointless, since his face, disfigured like the woman's was before, immediately appears right in the middle of the membrane. Aware that, once she's been found, it would look bad to pretend she hasn't seen him, she faces up to him and, putting her hands together, begs him both to help her and not to get angry.

Leo is very nervous, and he doesn't stop peering through the door, left and right. It's like he's looking for someone else and Celia, disappointed, instinctively steps back so as not to block his line of sight. Only afterward, from his gestures, does she interpret that he needed to measure the size of the platform, as he intends to come and get her in an aero'car. Him! He's coming to rescue her! This stuns her so much that, before she thinks to warn him that someone else is also coming to rescue her, the young man has already taken off. What a mess! Xis' mother getting her out of here would have had the advantage that Leo wouldn't have found out, but getting out with him ... is a dream come true. As good as it'd be to have either one or the other, both options together is a total nightmare. Who's going to get here first? If it's the doctor, she'll try to convince her to get her daughter out and leave her behind. Maybe if she explains Lu is on her way that'll work.

Unfortunately, they both arrive at the same time. The aero'car hasn't finished parking yet when Xis' mother unlocks the security system and bursts onto the platform like a woman possessed:

"Let's get out of here as fast as we can, kid."

"Thanks for coming to get me," Celia says, trying to stall her. "Where's Xis?"

"Don't even mention that. No one can know she's been at CraftER, do you understand?" She grabs her arm so hard that it hurts as she drags her toward the opening.

Celia fights back. With or without black glasses, she still really dislikes this woman. She doesn't understand why she's talking to her in such an intimidating way. Xis must be really miserable with a mother like this.

Just as she tries to escape, the doctor literally throws herself at her and pushes her forward without a second thought, but luckily Leo rushes out of the aero'car and stops them. Behind him, she sees Silvana and Lu run out looking very alarmed and, in an unthinking impulse that later she won't know how to explain and will make her blush, she also starts running and hugs not her mother but her tutor, tightly. All the accumulated tension is let loose at once, making her shake all over.

For Silvana, each wave of shivers feels more and more pleasant, and the human warmth spreads through her body until her eyes and her mind cloud over. Her wise hands take themselves to the neuralgic points on Celia's back where the shared memory of good vibes from the other day makes a space for the two of them, away from the others' reactions.

The long silence, which for some is very uncomfortable, is broken by Sus in the end:

"I can see you're all caught up on what has happened, and I hope that you'll be able to respect my wishes: all this took place outside CraftER, understood?" Her eyes burn a hole in Lu, searching for an impossible agreement, because it's clear that she hasn't even heard her, still shocked by what she's just seen. "That would be the icing on the cake if, as well as losing my daughter, I got fired... and maybe you too, Mar'10." She's changed tack and is now looking for compliance from someone else.

"You're the most affected. For everyone else"—Leo glances around at everyone, his eyes settling on Celia for a couple of seconds longer—"what does it matter where it happened? The facts are the same whether they happened inside or outside CraftER. No need to worry, I won't say anything. The only problem is the recordings..."

"I have them and no one will see them." Having taken care of the problem inside the company, she needs to make sure there'll be silence from outside as well, and she addresses the two women acrimoniously. "I imagine you don't want me to sue the girl. So I won't, as long as you never mention CraftER in connection with any of this. Erase that name from your minds, understood? The girls escaped from school and had an accident in the street. That's all Fi needs to know"—she looks at Lu insistently—"and everyone else. With the disappearance resolved, and with my consent as the mother of the victim, the POLis will close the case. If there are any leaks and someone defames the company"—her eyes are like daggers—"I will throw all CraftER's resources behind coming after you."

The threat seems out of place to all of them, even though Leo understands the motives behind it better than anyone. All the same the two women, after a brief exchange, and out of respect, they say, to the horrible predicament she's found herself in with the loss of her daughter, comply with her wishes. Celia is the most disconcerted, as she's interpreted "lost" in the literal sense of not finding Xis after school, and doesn't understand all this about suing and the accident down in the street. Taking refuge behind Lu, in an attempt to make up for her earlier effusiveness toward Silvana by returning her adoptive mother to her protector role, Celia keeps her mouth shut so as not to attract the rage of the doctor, who, it seems, is becoming a worse and worse person.

Certain she's achieved her aim, Sus Cal'Vin puts her hand in the cavity, and without even looking back as she walks off the platform with great dignity, she informs them:

"If, as I expect, over the next few days there is no change, I will clear the way for you to continue your research and will pass on positive reports to Dr. Craft."

Pure blackmail, Leo thinks, while at the same time pretending not to notice her tacit expectation that he follow her inside. He has Celia within reach and he doesn't intend to waste an opportunity to set a new date for an interview. Since he found out that the girls had snuck into CraftER he's been playing with the idea that Lu unilaterally canceled the interview and wants to confirm his suspicion. Maybe the anti-techno tutor and her pernicious influence have something to do with it.

Like him, the women and the girl have secrets to uncover, but nobody dares to expose them in public, and each one is speculating about the possibility of having some time alone with their desired person. Everyone except Lu, who for once has the upper hand:

“The girl needs to rest,” she says taking her by the shoulders protectively, with the intention of leading her to the aero’car.

Leo comes over very quickly and, taking advantage of the fact that he’d come up there with them, makes an offer:

“I’ll accompany you downstairs,” he says, gently taking Celia’s hands to help her inside.

The fondness with which the little girl looks at the boy isn’t missed by Silvana, as he repeats the gesture with Lu before turning to face her. The brief contact with his pair of wooden hands reminds her how long it’s been since she’s had any kind of relation with a pro-techno. The last one was Jul, her greatest failure. It was because of him that she is convinced the barrier is insurmountable: the wood never turns back into flesh. She looks at Leo and feels pity for him, and, consequently, for Celia. They’re worlds apart.

On the way down to the parking area, Lu takes the situation in an unexpected direction:

“ROBul, order her an aero’taxi,” she says, nodding toward Silvana. “We need to go straight home.”

This takes everyone by surprise. Especially Celia, who’d already accepted that she would have to temporarily say goodbye to Leo, if she wanted to enjoy a massage that would alleviate the profound unease she’s feeling. Petrified, she doesn’t dare snub her mother a second time, and limits herself to giving Silvana an imploring look.

“It would be a good idea for her not to go to school tomorrow,” is her immediate reaction. “Give her a sleeping pill and let her sleep for as long as she can.” Without saying so, she wants to make it clear that Lu mustn’t talk to her about Xis. “You don’t need to worry either: I’ll come over bright and early tomorrow morning to help her digest what’s happened. We’ll do an extra session, okay?” Now she’s talking to Celia despite the fact that she’s looking at Lu, and what she catches out of the corner of her eye doesn’t disappoint.

“Excuse me.” Leo interrupts the conversation as if he’s been hoping to intervene for a while. “I have an aero’car available. There’s no need to call an aero’taxi; if you want, I can take you wherever you’d like to go.” Any ploy will do if it can bring him closer to the girl’s world, and without a doubt the tutor plays a privileged role in it.

Silvana wasn't expecting this solution from a pro-techno. They never stray from their path, no matter what. In a moment of naivety, she even feels flattered, before she begins to suspect that he must want to get something out of this. What does it matter, right now she feels like taking part in a battle that she hasn't fought for a long time, against an opponent that reminds her of Jul. Maybe it'll be a chance to get revenge for the earlier defeat.

While they work out the details, Lu almost literally chucks them out of the vehicle. Leo, at last, with Celia's support and Silvana's unexpected silence, manages to get a promise out of her to return next week to complete the pending interview.

25

Despite not being at all happy about it, Silvana is obliged to let ROBco pilot. "It will be much safer than with me at the controls, believe me, it's a new device and I haven't piloted it yet," Leo confessed. She'd imagined she'd be alone with the boy, and the robot's presence is making her uncomfortable on a deeper level than what's implied by her principles. As much as she tries to hide it, she knows full well that the rejection comes from further down, from having resuscitated Jul's image, and with it comes a desire to relive that dialectical duel with a pro-techno. It bothers her that she has to speak in front of a tin man, and never has the phrase been more accurate, as the robot is right behind them and she feels like she's being spied on without even seeing the thing. The tempting fantasy of half an hour of excitement on the front line is about to disappear in a puff of smoke.

However, she also wasn't counting on her opponent playing the same game, and Leo is determined to reap some benefits from the journey. He won't give in to her silence so easily.

"If you'll allow me to ask a question... Why is it so hard for you people to accept that machines can perform some tasks better than we can?"

He's leaning his head toward her in such a condescending manner that Silvana feels an urge to punch him. What does this fool take her for? She visibly turns around as if she were looking for someone behind her before letting fly:

"Are you speaking to me in the plural or does your ROB not accept that it drives better than … us?" The irony is pouring out of her eyes. "And, by the way, who are 'we'?"

Leo is taken aback. He'd imagined she didn't like him very much, but he'd hoped the offer of taking her home would improve his circumstances. He's especially dismayed by the tone she's employed, so far from the formality she'd used with him up until now.

"Sorry if I've offended you, it wasn't at all my intention. Believe me, I totally respect your point of view."

"And what do you know about my point of view?"

"The anti-techno view, I mean." He pauses, as if he doesn't dare give voice to the thought that's playing in his head. "You know … you're the first one I've met in person."

Here we go. Good thing he's given up the fight … but an opportunity like this can't be wasted.

"And what? Do I seem very eccentric to you?" She turns to face him with such force that she hurts herself on the seat's ergonomic springs. This cutting-edge comfort technology is all very well, but the designer never even anticipated that passengers might want to talk to each other.

"No …" He limits himself to turning his head, a movement that is completely appropriate to the environment and that Silvana considers typical of a ROB. "Actually, while we were looking for Celia it didn't even occur to me; it was seeing how the girl hugged you that reminded me." He finishes the phrase slowly, as if he were still getting his head around the emotional impact of the scene.

Now she's the one who feels lost. She'd been expecting a counterattack from the technological side, disdain for her withdrawal from progress from the boy, following the strict pro-techno canon, but in the end he's brought up Celia and the hug. Jul never let her take the fight into this territory, for him such things were mere trifles and it was more important to focus on the great challenges that would shape the future.

"Maybe you'd never seen a hug before." She could lace this comment with sarcasm, but she decides against it.

"Not like that one."

The boy's honesty is disarming. She'd never experienced a hug like that one either, and less still in the flesh. Her hair stands on end just thinking about it.

"She's an extraordinary girl. I won't allow you to make her suffer for CraftER's benefit."

"That's what I thought: it was you who advised against her coming to the interview. And, the truth is, I don't understand. I agree that Celia is a prodigious child: that's why I need her for my project. She's delighted about participating in it, and it can't cause her any harm."

"Don't you think you've done her enough harm already? Losing a friend in those circumstances can traumatize a person for life. A century ago, the grieving period lasted years, a whole life even. It's impossible for you to understand, even I can't get my head round it, and I've studied it."

"Hold on a minute, let's take this one thing at a time. Them being at CraftER today had nothing to do with the project. And the girl didn't seem very affected when she insisted on setting a date for the interview."

"Her behavior seemed normal to you, I can tell. Well, it's obvious she doesn't know her friend went over the edge, it doesn't take a genius to see that. And if she hadn't had an access code, none of this would have happened."

"I'll be more careful next time, I'll cancel her code when she leaves, but don't stop her from coming." His tone is almost pleading. "It might even distract her from the trauma you say she'll experience; the other day she had a good time, ask her. And the recordings are harmless."

"How can you be so frivolous? You don't even know if working for that company might be detrimental to you. What are you trying to do now? Making robots with feelings … and you have to suck them out of a little girl?"

"No, no, please. I tried to explain it to you the day of the get-together: it's about boosting human creativity"—he hopes saying it like that will make it sound better—"by giving people an assistant that stimulates them."

"Very nice … but do you believe in it?"

Suddenly a neutral voice interrupts the conversation:

"*Information*: I am an example."

The shock makes Silvana jump and stab herself on the springs again. Absorbed in the conversation, she'd forgotten they had a silent witness.

"What is it saying? That what you took out of Celia you put into this thing?" she shouts, pointing at the robot, one wrong move away from destroying her back in the process.

Leo is about to lose his patience. He rues the moment he decided to take her home. Instead of winning her trust so she'd let the girl come to CraftER, he's achieving just the opposite.

"Calm down, please. Nobody's hurt Celia. What ROBco means is that it has been fitted with a prototype of the prosthesis we're developing. You see, I'm the guinea pig, not the girl," he concludes, with resignation.

Poor naive boy, Silvana thinks, it's quite possible that is the case.

"And you're okay with that ... ?"

"Yes, think about it: it's a device designed by me that helps expand my capabilities. What more could I want?" He never would have thought he'd end up defending the Doctor's project so convincingly.

"Machines that augment human capabilities seem like a great idea to me: without remote manipulators surgeons couldn't operate on a microscopic scale and, without INFerrers, we'd take too long overthinking the consequences of our decisions ... it's ROBs that I reject, and the personal link that is established between them and their PROPs that ends up hogging people's most intimate time and space. You said it yourself: you don't need anything else ... and, in the end, you become wooden like them."

"That's what really gets me about the anti-techno lot"—Leo can't take this anymore—"you confuse everything, you get it all mixed up. First off, I was talking about expanding capabilities, not augmenting them. The machines you're so fond of are useful, sure, but they only magnify what we already have. I'm talking about creating new skills, broadening the range of what we can do. For example, ROBco ..."

It's the first time Silvana is actually listening to him and she must admit the boy is really good at splitting hairs, in that respect he's just like Jul.

"*Question*: Would you like a suggestion?" Upon receiving Leo's assent, it goes on. "Try not to repeat yourself. I have already been used as an example and it is obvious that she does not like ROBs. Look for another example, one that appeals to her more."

"Don't you find it degrading when he talks to you like that?"

"Why? He's given me some good advice. Quite the opposite, I'm pleased the prosthesis is working."

Without a doubt this idiot is as wooden inside as he is on the outside. Now he'll make an effort to obey the robot.

"Let's see. What do you think interests me?"

"Celia, or more precisely, her feelings; before you were afraid we were sucking them out leaving you empty-handed, I suppose." Leo couldn't possibly imagine how right he is. "What would you like to broaden in yourself? Maybe your repertoire of emotions?"

From Silvana's face he can see he's hit the nail on the head, while she still can't make sense of how irresponsible people can come up with such good results.

"Great, sure, how would you go about recovering extinct emotions?"

"If they're extinct they must have existed at some time. Now I understand: you're interested in the girl because you can't find these emotions nowadays. But I was talking about new feelings, that have never existed."

"Alright then, how would you create them?"

"Off the top of my head ... I don't know." His brain is working a hundred miles an hour, he needs to make the most of this opportunity. "Perhaps the closest thing I can think of is that ... sentiments are like a color palette: grief is black, rage is red, serenity blue ... by combining the basic colors, we can obtain many more, I guess you must know if a set of basic emotions has been identified, and then ..."

"Yes, of course, there are seven. But not all of them are compatible nor can they all be combined ..."

"It was just an analogy. For me feelings are activation patterns in the nervous system and, as such, can be added together, subtracted ... the important thing is to have basic patterns at your disposal that cover all emotional dimensions." Spurred on by the woman's direct and insistent look, which is becoming less aggressive, he continues. "Maybe that's what's missing, maybe we're only operating within a restricted subspace of colors: the blue-red strip, let's say, and what we need is to discover yellow."

"A new dimension, I was saying that to a colleague just the other day: the problem lies in finding it."

"It could be attempted with intracranial stimulation devices, though the number of possibilities is so high that ..." He cuts himself off suddenly. "I can see bioengineering is starting to interest you a little more." They've landed in front of the ComU and Silvana is making no move to exit the vehicle. "Before you go, I would like to ask you a question, too: do you know why the girls escaped and headed over to CraftER?"

"Look, it's a creative action: no one escapes from school nowadays. You should know that, it's your specialty."

"Was it Dr. Cal'Vin's daughter's idea or Celia's?"

"You took recordings of her brain, right?" She places her hand flat against his chest as she often does with friends when they're arguing about something, but when Leo jerks back she immediately removes it. "So do an in-depth analysis of them and maybe you'll find the answer."

"If I find it, do you want me to tell you?"

"Here's a personal connector." Leo is taken aback, first because she uses the thing and second because she's giving it to him. "Anything to do with Celia interests me, and if you come up with a means of creating emotions, let me know as well."

Just as during the trip both would have agreed to cut it short, now that it's finished, they'd make it longer. While one is wondering what he'd have to add to ROBco so it could adapt the driving speed to his whim, the other is saying goodbye and encouraging him to use the connector soon. If the aero'car's windows weren't opaque, when turning to watch it take off Silvana would be surprised to see Leo with his hand on his chest in the same spot she'd placed hers. And she wouldn't know what to make of it.

V
TRANSMISSION OF SENTIMENT

26

Pain in its purest form.

This turn of phrase, with its surgical precision, is what Silvana would use to diagnose the shock Celia is suffering from if she were to take her as an object of study, if she were to consider her symptoms and go on to catalog them as part of her collection of samples. She came prepared to explain Xis' death to her, tactfully, very tactfully, massaging her tender body's neuralgic points in order to reduce the impact of her words. And then she would accompany her throughout the grieving process; she's been working on it all night. The girl needs help to work through things: to receive the information, digest it, and eject it. Three stages that a hundred years ago took months, and that nowadays, in the rare cases that they do occur, are over in a matter of hours.

However, upon entering the bedroom and seeing Celia lying prostrate on the bed, still wearing yesterday's clothes, with the saddest expression she's ever seen and her eyes half closed, blank, not even registering her entrance, Silvana's legs give out and she has to sit down. Near the head of the bed the girl's robot doesn't take its eyes off her.

Knocked down onto the chair, she perceives Lu's pout, and it's not difficult to imagine why she's wearing it: the professional is failing her. As she struggles to pull herself together, she realizes someone must have told the girl what happened to Xis, it's the only way she can account for this drastic mood change. And that someone is unlikely to be the robot, too inclined to follow orders, so it must have been the stuck-up woman standing before her, who, even still, manages to look at Silvana with disdain. She doesn't know why she hates her more: because she ignored her advice or because of the pain she's inflicted on the little girl.

"Why did you have to tell her?" she whispers out of the corner of her mouth so Celia doesn't hear her.

"You have no idea how difficult she can be. She pestered ROBbie until his alarms went off and ROBul was forced to wake me up ... at three in the morning! I authorized him to tell her everything, yes, so what? You would have told her this morning anyway, right?"

"Of course, but ... "It's not worth the effort of trying to get her to understand the difference. "How did she react then? Did she cry?"

"I don't know, when I got up she was like this. I think we should tell the clinic. She was frozen for such a long time, poor thing, maybe it's normal for her to become lethargic from time to time. They should have warned me though."

There's not much to gain from talking to this airhead, Silvana thinks, while she instinctively turns toward Celia, unable to figure out whether or not the girl has heard all this nonsense. Reluctantly, she admits to herself that even the robot would provide more information, although, at the moment, she won't even think about lowering herself to speak with it; she'll keep that as a last resort.

"Could you leave us alone? I'd like to do the massage the same as every day."

Lu's face doesn't hide her satisfaction that, even when she wasn't expecting it, the professional has taken on the problem, and she dutifully leaves, followed by ROBbie. If it weren't Celia lying on the bed, Silvana would quickly have made it clear that since they ignored her instructions, she can't make any guarantees.

Celia's body, lax, doesn't move even a millimeter as Silvana approaches: her knees are slightly bent, her hand dangling over the side of the bed, almost touching the floor, and her eyes are just as absent as before, despite the fact that, for a moment, she had convinced herself they were following her. She sits down next to her feet and strokes them softly, taking extreme care not to annoy her. There's no reaction, either for or against.

She knows full well that she must be patient and wait for the stimulus to make its way through, breaking down barriers and rebuilding ruined bridges.

She keeps going and going, until she notices that the feet give themselves over to her and, when she looks up, so do the girl's eyes. The stabbing pain is so intense that it smothers everything else, so much so that she's certain she'll never forget this moment. Silence, she's better off focusing on physical contact, she wouldn't want words to impede their understanding, fragile as it is. Without losing visual contact, Silvana's dexterous hands work their way up Celia's scrunched up legs until they pass her hips and, traversing the profile of her thorax, they ride up over her back and finally take control of the base of her neck. She can feel that the ball of muscles is stiff with tension, so, extremely carefully, she devotes herself to unknotting it and relaxing them one by one, conscious that it's a delicate situation, and it'll be even more delicate when she stops and it's time to talk. It won't be her who precipitates the massage coming to an end.

It's been a while since Celia lost touch with the person she was before and, freed from the deep well of thought, she has been transformed into just gazing eyes and skin. No more. The sight of a friendly face infuses her with bodily well-being, and she's no longer herself, but a cadence, a coming and going of waves that rock her and lift her and turn her into a spark at the mercy of the current. Until, all of a sudden, pampered by the magnetic blue eyes that are so close to her, she sits up and surprises Silvana with an even more intense embrace than yesterday's at CraftER. And also wetter, because the tears she's been holding in all night burst out with all the force of a great storm.

It's like time has stopped, as neither of them makes any attempt to untangle herself from the other. And once again it's Celia who, when she stops sobbing, whispers to Silvana while still holding her:

"I want to die and be with my parents"—before her voice breaks because the tears have returned.

"Don't speak, sweetie, and you'll start feeling better, you'll see." She hugs her even harder against her chest, but when she detects resistance, she separates herself a bit and finds a cry of despair in Celia's eyes.

"You want me to shut up too?"

"Of course not, you know I love it when you tell me what's going on. I just thought that today it would hurt you."

"You know, don't you? You all know but ROBbie had to tell me." She looks away, like she's withdrawing her trust.

Silvana takes her hand, she needs to maintain contact.

"What exactly did it tell you?"

"Are you afraid of telling me more than I already know?" Her belligerence has dried her tears.

"It's not that …"—she chooses her words carefully—"but I would prefer that they hadn't told you anything."

"And then what? Would you have taken me out of school so I wouldn't find out?"

"No, no and no!" She's practically shouting, holding onto her shoulders. "I want what's best for you, you know that right?" She gently shakes her and tries determinedly to look into her eyes.

"It's my fault Xis died." Silvana has prepared so much for this moment, but now she can't even breathe. "I should have jumped off after her … what am I doing here? Suffering and ruining everything. I would've been better off if the tumor had killed me."

Silence falls heavy as a rock between them, and Celia doesn't receive the soothing response she yearns for, that she needs, that she would beg on her knees for if it would help … and Silvana's not capable of giving it to her. Now it's her who's been transplanted into another century and here massages won't help. It takes her an eternity to react, and in the end, the words come out reluctant and forced:

"Your mother … What would she do? What would she say?" Her eyes are shining, and an emptiness has opened up inside her, a new space that's devouring her. "How would she make you feel better?"

Celia's hand automatically heads for her pocket and takes out the ring. She'd been so distraught that it hadn't even occurred to her to use it, and now she's looking at it as if it were an apparition, so absorbed that Silvana's question makes her jump:

"Was that your mother's?" The silence makes her think it was. "Put it on, it'll give you strength, she always knew how to make you feel strong."

Celia's brusque movement catches her unawares.

"It's too big for me, you put it on …"—she thinks it over for a moment—"and then talk to me."

Her words are as transparent as her back muscles, Silvana thinks, while her shaking finger welcomes the treasure offered up to it; even though she's from another century, she's still a little girl. This thought calms her a bit and, nervously touching the ring that for years rubbed up against a skin that Celia loved so much, she searches for inspiration about how to talk to her in the way she expects.

"You know I'm very fond of you." Her voice resounds with a solemnity she would prefer to avoid. "I'm not like your mother, I don't have the same type of brain, and I haven't lived through the same experiences. I'm from another age, I have a different way of doing things than hers ... and yours." She's getting tied up in knots, she should be more direct. "But if you guide me and help me, I'll try to make it so you can speak to me like you would to her."

There's no physical contact between them, only visual and, incredibly, Silvana doesn't miss it. The electricity coming from Celia's sad, yet hopeful eyes is enough, as it prickles through her whole body, all the way to the very deepest parts.

"I already told you: it's my fault Xis died. I persuaded her to escape, she was really scared, and when she needed me most ..." Enormous tears, slower and steadier than those that came before, slide down her cheeks.

"Calm down, my love, it's not your fault." Silvana's hand can no longer resist trying to dry the undryable, and the ring makes its way across Celia's face in a long caress.

"I must have fallen asleep ... she was so anxious, and I ... I could have tried to hold on to her, or talk to her ..."

"Don't torture yourself anymore: the cause of this sickness is part of her, not you. It's that damned education the pro-technos get: without their ROBs they feel completely lost, helpless. A child from the ComU would never have thrown herself off."

"I should have known. I never should have separated her from ROBix. And I took her up there ... how awful." She starts blinking rapidly in an attempt to hold back the tears. "And her mother ... I understand why she treated me that way. What am I going to do now?"

"Don't worry, Sus Cal'Vin won't do or say anything, it's not in her interest."

"What do you mean?"

"She made it very clear she wanted silence. And, if no one makes a claim, the POLis won't investigate. Case closed."

"So nothing's going to happen? No one cares how she died?"

"Look, it was an accident, that's all."

"But, if it weren't for me, Xis would still be alive. She saw the danger: she told me I was from another century and I had no idea how things work. I didn't listen to her and she was totally right: I have no business being here, I'd have been better off dying when my parents did."

"It hurts me to hear you say that … Can you imagine how your mother would have felt? She wanted you to live. She knew you were a great person and you would do great things. Didn't you tell me that's what she wrote in her letter?"

"Yes, but everything I've done has been bad."

"Not at all. You've been good for Lu, for me and, it seems, at CraftER they're over the moon about your contribution to the project." She's said this as a last resort, conscious that it's a powerful weapon, though she never could have imagined how powerful: Celia's face is transformed in an instant.

"Who told you that?"

"Leo took me home last night, remember?" Now she really has her, all the girl's attention is focused on what she's saying. "He's extremely interested in taking some more recordings from you—he's convinced he'll never find anyone more creative."

"Really? But you don't like the project."

"I didn't like it, no, but when he explained it to me I started to see it differently … or maybe it's this ring that gives me a different point of view. I'm sure your mother would be very proud of you for taking part in it."

"My dad more. He would be so excited to go into CraftER and see all the machines. Things would be so different if he were here!" Another blast of homesickness fills her eyes with water once again. "And you, would you want to go there with me?"

"If Lu doesn't mind …"

Lu, another problem. Celia finds it impossible to talk to her, she's afraid she'll make her go back to school.

"Please, please," she begs insistently, "let me go to the ComU every morning, we can do the sessions there at least until the CraftER interview."

And then … what then?, Silvana can't help asking herself, conscious of the danger it would pose to the girl to live only looking forward to one day and one time. She knows she's thinking too far ahead, but what choice does she have aside from letting Celia cling to a hope that, before she's even thought about it, has already taken hold inside of her as well.

Discussing how she'll suggest it to Lu, the time continues to plod on steadily, with neither shocks nor risks: nothing more than jumping from stone to stone to avoid the vertiginous nature of the deep hollows. It's only when it's time to go and confront her adoptive mother that Celia wavers, about to fall back into darkness. But Silvana has already claimed the territory that she now treads with relative confidence:

"Here, keep hold of this extraordinary ring," she says, handing it over cautiously as if it were made of glass, "but you'll have to lend it to me again, okay? For a little while in each session ... so we can have a conversation like today's. I've got a lot to learn, about you, about myself ... and about her," she concludes, reverently observing the ring that's already slipped into Celia's pocket.

27

Ever since Silvana said all that about analyzing Celia's recordings in depth, that maybe then he would realize what was going on, Leo has been dedicating all the free time he has left over from the E-Creative project to studying them and trying to find their hidden meaning. With no success. The indicators he's extracted are as unequivocal when it comes to corroborating the girl's creative talent as they are useless for working out why she snuck into CraftER.

As a last resort he's thought of injecting the signals into his own brain using the sensory booth. The problem being that the booth is at home and the records can't be removed from the company building. In the hours prior to the delivery of the second prototype he began to hope that the Doctor would be so satisfied with it that he would authorize the exception. He would be forced to admit that Leo had produced something of a high-water mark and, ultimately, the volume of information that he wanted to take home with him was insignificant in relation to the ridiculous amount of gigabytes that would make up the prosthesis. But when it came to the moment of truth the same thing happened as the last time, he really can be naive: the new prototype has been installed in ROBco and he will once again be the guinea pig.

He's had no choice but to take apart the key parts of the booth, transport them and try out the transmutation right here, in his cubicle. He's got plenty of space. The only danger is the Doctor confiscating his invention, and then not letting him take it out of the CraftER building; but he'll have to take that risk if he wants to respond to Silvana. And he does want to, even though his brain is denying him any convincing excuses. It bewilders him that it's an anti-techno, of all people, who has pushed him to investigate technological tools that, according to her principles, she rejects, and he justifies his own desire to please her by citing the priceless revelations he'll receive from her in exchange. Revelations about Celia, the project's most valuable subject, he tries to convince himself, while privately he knows well enough that it doesn't explain the burning sensation he feels in his chest every time he thinks of that hand she placed on his body.

With ROBco's invaluable help, he's reconstructed the booth in no time at all, irrefutable proof that the learning module is working. How different from that idiot he had at home, which he had to ban from coming anywhere near his devices while he was using them. He can't begin to imagine what would happen if humans progressed at this rate. For a moment he feeds the fantasy that this is what the Doctor is after: fed up with improving robots, he's decided it's the humans' turn; but he wouldn't believe that even at his most suggestible: at best Dr. Craft is trying to improve himself to increase his domination. And Leo is contributing to that. In return, he must admit, he also benefits: he has the privilege of having ROBco available to him and, by a twist of fate, now it's him who receives all the advantages of the prosthesis. The downside is that, when the project ends, he has no idea what will become of his efficient collaborator, most likely it'll be mutilated in order to reduce it to its previous state. He'd better get thinking, and quick, if he doesn't want to end up with a rudimentary servant once again. And that wouldn't be the worst of it. It's better not to think about what will become of himself.

Almost as if it senses his fear and wants to assure Leo of his survival, the robot is being more diligent than ever today. While the boy was lost in pointless worrying, it has applied the basic tests to check that no device has been damaged during its transportation and that they've been reconnected properly. It has also turned on the recording devices just as it's been ordered, because Leo doesn't want to miss any details and isn't sure he'll retain any memory when under the effects of the signals.

"*Information*: Everything is prepared to initiate the experiment."

He's more nervous than he thought he'd be. For no reason at all, since he can inject himself with Celia's signals as many times as he likes. The colossal Kun Yang's even bored him in the end, but that was different: as impressive as the jumps and slam dunks were, everything was reduced to muscular sensation and seeing things from unexpected points of view, while today he will feel like an intruder hidden behind the door of the girl's most private room.

He orders ROBco to adjust his helmet, and closes his eyes to avoid any interference. The waiting period is composed of a blackness speckled with red spots that seems to go on forever, until a buzzing in the right side of his brain tells him it's beginning to work. For a good while he remains conscious, maintaining a separation between himself and the outside influence coming in. He can clearly see the unfinished drawings he projected to Celia, and hugely enjoys the flashes of their possible completions, some of which are really surprising, while others are impossible for him to interpret. Like the object with two wheels that looks like an old vehicle cut in half, or the long stuffed toy with more legs than the nanorobots that clean his arteries. The continuous flux of forms that go round and round his visual cortex without the slightest effort conveys a certain tranquility that makes him drop his guard, and the cameras and sensors installed in the chair record how his body becomes more and more relaxed.

That is until the drawings are finished and the 3D objects sequence begins. It makes him uncomfortable to see simple tools subjected to such extravagant uses, like the poor control lever that gets longer and longer until it touches the floor, or the opposite, it shrinks to a ridiculous size until it becomes pointy and sharp; or it's pushed around by the wrong end and the handle is used to hit a sphere around the floor, and it's even turned into a thin stick that is inserted into the wall, and then is thrown up in the air with many other sticks, only to roll around the floor once more, having become fat again. Although he doesn't notice, Leo is sweating, and his muscles have tensed again. It's not so much because of the strangeness of what he is viewing, instead it's due to the huge amount of energy that is sucked out of him with the creation of each image. So much different than before. A shadow of pain extends through his whole body and, instead of trying to escape, he feels attracted to it. The abyss pulls him in with an unnatural force, and he doesn't know where it will take him, when, all of a sudden, an inoffensive sketch of fingers laced together hurts him deep inside, and his eyes, wide open, are dragged toward his own hand, lying inert on his thigh. He can feel a strange warmth there, as if his fingers wanted to stroke the palm that holds them together, in the same way that his lips are burning with the desire to float over there and kiss that hand or those fingers, melt into them, lick them.

Captivated by such a disturbing sensation, Leo has taken the final step and has forgotten himself completely. His brain, drowning in Celia's feelings, adopts patterns of activity that until now have lain inert, and sends brand new nerve signals to the other organs in his body: his lungs take in less air than usual and his muscles weaken to favor his heart, which is beating wildly and pumping more blood than necessary to his face.

Who knows what might have transpired if ROBco hadn't intervened. Upon recording such consistently unsettled vital signs, there was no doubt it had to apply the emergency protocol to gradually cancel the session. To do it suddenly might have caused irreparable damage to its PROP.

Leo takes a while to come back to himself and, once he has, he looks around him, astonished: he doesn't know where he is or what has happened, and when he sees ROBco so close by, the shock makes him jump, reminding him he's wearing the helmet. Of course, the helmet, the experiment. He goes to stand up without knowing why, and ROBco has to stop him.

"*Warning*: I cannot allow you to stand up until your parameters have returned to their base values."

"What was that? Did I lose consciousness?"

"*Affirmative*."

"You mean I fainted and that's why I don't remember anything?"

"*Negative*."

"So what is it? Come on ROBco, don't break down right when I need you most. Explain yourself."

"*Information*: You did not faint, because your muscles maintained their tonicity, but you were not conscious. The injection of the signals altered your vitals and I had to abort the session. It will be necessary to analyze the recordings to find out exactly what happened to you."

"Will we find anything? They must be very short."

"*Specification*: thirty-three minutes and fourteen seconds."

Leo is stunned. The most he can remember is a couple of unfinished drawings and how their completion was imposed on him, out of nowhere.

But more than half an hour... Now he does have a good reason to get up and this time ROBco doesn't stop him.

Standing before the neurovisual player he can't believe what he's seeing: his face, placid to begin with, contracts more and more until it becomes unrecognizable. He's never seen himself looking so disturbed, and he has to thank ROBco for putting an end to such extremes of distress and excitement. He hadn't noticed Celia suffering from anything similar while he was recording her. When correlating the limbic activity with the retinal images, he realizes that the highest peak coincides with him observing his own hand, and he wonders if the transmutation can be so true to the original that the girl was looking at the same thing. How can simply seeing a hand cause so much fuss?, he asks himself as he looks at it in detail. It doesn't make sense, none that he can think of. He's in a real muddle: if the interview caused such an upset in Celia, he's even further from understanding why she wanted to come back to CraftER. He's got no answers for Silvana.

He's not a man who gives up easily, however, and, taking advantage of the fact that he has the prosthesis at his disposal, he turns to ROBco for help. Among the many questions the robot bombards him with, one in particular takes Leo by surprise.

"*Statement*: You keep speaking about the impression your hand might have made on Celia. I suggest that you symmetrize the thought: what impression did her hand make on you?"

He must admit, with surprise, that he has no memory, either visual or tactile, of the hand that he assumes is small, even though he did see and touch it. Leo, who attributes his success to a portentous memory, capable of reproducing information in the most minute detail. He would argue that the sensory world didn't matter to him if he couldn't still feel, very much real, the imprint that another hand had left on his chest. That had never happened to him before, but neither had any woman touched his chest, aside from during sexual gymnastics.

Ignoring Leo's silence, ROBco continues along the same lines:

"*Addition*: As well as symmetrizing, you could try generalizing and searching for analogies. I *remind* you that the same idea applied in different contexts is reinforced as a result."

"Stop, stop. It's pointless telling me so many things at once. You have to let me think between one piece of advice and the next."

"I am *reducing* the velocity_of_suggestion parameter by half. Do you approve?"

"No. You have to wait for a response before moving on to the next thing."

The confirmation that, thanks to the trick of installing the prosthesis in ROBco, the Doctor ends up profiting from all his actions, shoots through Leo's mind like a rocket. He pushes himself to stay focused on the train of thought that was opening up an unexpected perspective for him, the one about hands and different contexts: could it be possible that Celia wanted to see him, just like he wants to see Silvana? The illusion doesn't last even five seconds, he might as well get it out of his head right away: there hadn't been a single moment when he had wanted to touch the girl, whereas if it weren't for the touch of that other hand, right now he wouldn't be caught up in these digressions. It must be something else... and that feeling of suffocation, that anxiety... Without being plugged into the machine, he can't feel any of that.

He plays the recordings back again, and it must be one thing or the other: either the transmutation is faithful and the girl is capable of feeling things in absence of any kind of contact and without letting anything show, or he's going to have to revise his invention, because it causes unpleasant side effects. It's a shame he wasn't able to record what he was going through when he was making that face, because he's sure it was an innovative experience; but, although he likes a bit of adventure, he's not prepared to go through that predicament again.

An innovative experience—the expression is going around and around in his head—isn't that what Silvana was looking for? Creating new emotions... not quite, she called them something different: extinct, she said. It makes no difference whether the unease he suffered was anachronistic or an unwanted side effect of a defective transmutation... he's sure she'll be interested, it's not like any existing emotion. And she was very clear that, aside from anything to do with Celia, he should inform her about any discovery of new feelings. Now he can use the personal connector without being embarrassed about showing up empty handed, and who knows, she might be able to help him untangle the knot in his recordings. Since she dedicates herself to the study of emotional states, surely she can detect and classify the one in there.

He turns to ROBco to thank it for the last suggestion and ask it to activate the connector, and finds that it's left his side without warning him. Once he's over the initial strangeness, he realizes that he'd removed any possibility for it to interrupt him and, of course, after a period with no response, other processes must have taken priority. He'll never manage to get it right.

Once he calls it, though, connection is immediately reestablished. Before he's even had time to think about what he's going to say, Silvana appears on the monitor and takes the initiative:

"I thought you'd never get in touch. Does it take this long to process a few signals?"

"Noo no ...," he stutters. "I've been snowed under finishing a prototype." How can it be that he finds talking with this woman more daunting than answering to Dr. Craft? "But I've made a discovery that might interest you."

He could never even begin to image how much. The enthusiastic reaction he receives when he explains it to her so surpasses his expectations that Silvana's offer to try out the transmutation herself takes him completely by surprise. How can she say that now, if she was so against letting him record Celia? For a moment his wires are crossed and he wonders if he's confusing her with the mother, but he soon regains control:

"I'll have to get an entry permit. And I must warn you that it might be unpleasant: in the recordings it's clear, as I told you, that I'm pretty distressed."

"That must have been the anxiety the girl felt about striving to give creative answers in the test."

"No, it was different. I also perform creative tasks and I've never felt like that."

"You told me you play chess, right? So you must have noticed that when making critical moves just before a win you sweat more and your pulse and respiration accelerate."

"Why do you say before winning? It's when I'm about to lose that I get anxious ... but that's nothing compared to what I can see in these images of myself."

"Everyone thinks that the one who gets worked up is the loser. But it's been demonstrated that just the opposite is true: a certain level of anxiety stimulates your mind."

"I'm telling you it was really different, it felt more like suffering, like when you experience physical pain."

"In the old days women knew full well that to procreate they had to suffer."

"Are you comparing the pain of childbirth with what Celia was feeling when she looked at..."—he's about to tell her about the hand, but he corrects himself at the last moment—"I mean, when she was looking at the pictures?"

"Let's see, I don't know, anxiety can be caused by many things, often all mixed up. All I'm saying is that creation requires a certain discomfort; and it's curious to me that you of all people are surprised, since you're carrying out an investigation on that very subject."

"So you're not afraid of subjecting yourself to that discomfort?"

"Is there some problem that makes you want me to take it back?"

"No, quite the opposite." It's foolish of him to show reluctance when given the best opportunity he's had for days: he will receive a visit, of her own accord, from the woman he's been longing to speak to again, he has a volunteer for his experiments, and at the same time an expert in emotional states, all in one fell swoop. "I'd like to meet you before the interview with Celia, which we've set up for next Wednesday. Could you come this weekend?"

"Perfect, that way I won't have to make any changes to my work schedule."

They agree that he will let her know as soon as he acquires the permit, and when there is nothing left to arrange, Leo thinks to ask after the girl and whether they've told her that her friend threw herself off the platform yet.

"Poor thing, that really is suffering. I doubt that all your devices combined would be able to capture it."

A timely shudder seizes Leo, allowing him a glimpse of what it would be like to transmute himself into Celia right now.

28

Silvana feels strange, she's not used to hiding where she's going and why, though she's always claimed she doesn't have to explain herself to anyone. Luckily she's already in the aero'taxi and on the way to CraftER, without having raised any suspicions within the ComU. Baltasar being totally snowed under all week with the preparations for the annual gathering of ComUs made it easy for her; and when it came to Sebastian, she's avoided being alone with him the last few days, just in case. She's convinced that she has to keep quiet about it, but some part of her nature, which she's been suppressing, keeps nagging at her. See how foolish she's being: not only is she mixing with a pro-techno, like when she was with Jul, but she's also going right into his lair. No matter how much proof she might obtain of extinct emotions that might be recovered, no one at the Ideological Committee will ever approve of her having taken this step. Not even she would agree with it, if she were a member.

This is a good reason not to assume that kind of responsibility, Silvana tells herself fiercely: If she had to give up her sense of adventure, what would she have left? A long string of years stretched out before her, dedicated to the same old bodies and identical slogans, what a stimulating perspective. She'd always have old books to turn to, of course, and cases like Celia's... Celia. With her it's been difficult not to let slip anything she didn't want to say. Every time Leo has come up in conversation, and he's come up a lot over the last few days while they've been doing the sessions at the ComU, a spike of adrenaline has put her on high alert inside, though on the outside she's made an effort to maintain her voice and posture, so as not to give anything away, or, on the contrary, to come across as too inscrutable. It's been really difficult, as Silvana is moved when Celia confides in her, but she always has the sense that Celia keeps her most intimate thoughts to herself. Rarely in her entire life has Silvana felt so powerless. The more she learns about Celia's hidden depths the more it seems she has to discover. And the more she realizes, painful as it is to accept, that her massage is but a superficial tool, completely useless for any further excavation.

She wishes she could convince herself that the bioengineer, for whom Celia feels a strange affection, she's sure of that now, could help the girl in some way. Maybe by putting herself in the child's skin and experiencing her feelings, she will be able to help her too. She's not afraid of suffering if it's for a good cause, and Celia is the best cause she's had anywhere near her since she became affiliated with the ComU. Yesterday, when the tips of her fingers were manipulating those muscles that form part of a landscape she now knows so well, she felt a bit treacherous keeping today's appointment a secret, but on a conscious level she didn't know how to tell her. She has no desire to cause Celia any more harm than they all already have.

As soon as she begins to discern CraftER's shining facade, the aero'taxi starts its descent. Today it won't be landing on the eighth floor like last time, nor will she have to remain on the platform. Leo has gotten her a permit to enter not only the company, but also his cubicle, a privilege that not even Celia enjoyed but that would certainly make her happy.

Silhouetted against the brightness of the main entrance, a masculine figure begins to take shape as she gets closer. The backlighting, the angle of approach, the imminence of the unexpected as the figure gets closer and closer, everything comes together in a flash: a memory of Jul coming to find her, without warning, to give her a heads up that things weren't going well in the negotiations between her commune and his colleagues. It was the best result of a long and otherwise sterile mediation: confirming that she could seduce a pro-techno to the point that he would risk betraying his convictions. How moving, that gesture of loyalty that kept her company at her lowest moments. She can't believe she'd forgotten that. If it weren't for this combination of circumstances she might never have remembered, and her memories of Jul would have been reduced forever to the moments of hand-to-hand combat—both the rational and the not so rational—with that powerful yet inexperienced colossus that are so easy to evoke.

For Silvana it's an excellent sign that Leo has arrived early and is waiting for her outside. He's wearing a jacket similar to the other day, but this time it's a maroon color that suits him better, with the same horizontal, black stripe across his chest that highlights the width of his shoulders. She was already looking forward to the meeting, but now it feels even more exciting.

When she gets out of the aero'taxi, holding his hands, she feels an impulse to kiss him on the lips, but she suppresses it just in time. She must avoid any outpouring of emotion that could embarrass him; last time they met it was pretty obvious that the mere touch of her hand, outside of the established conventions, made him uncomfortable. They've come so far with technology, these pro-technos, she thinks, as she puts her hand into the opening the boy is showing her, but when it comes to physical contact, they haven't moved on since Jul's time. Perhaps they're even more ignorant: the third and fourth generations probably don't even know why they live here and us over there, what it is that separates us.

Neither the downstairs security system nor the mobile platforms that take them to Leo's cubicle surprise her much compared with what he explains about the timeout device. She's not at all pleased that, when she leaves, she'll forget part of what has happened, and, most of all, she fears there may be harmful side effects. Frozen before the device, she complains that he hadn't warned her when they made the appointment, and at the same time wonders whether she should back out.

"I was so surprised you offered to come that I didn't think to warn you about it. But there's no need to worry: I can guarantee it's totally harmless. I've been in and out loads of times."

"And has your brain been wiped every time?"

"It's not wiped, the information is still in there, but you can't access it."

"How terrible to be denied access to a part of oneself."

"Actually, it's not that strange or artificial. Memory works like that anyway, you must know that. Often an old object or specific visual or auditory surroundings can recall a memory that, otherwise, we never would have recovered."

The boy has a gift for this, Silvana thinks: the other day, despite his total ignorance, he guessed that she longed to discover extinct emotions, and now he's hit the nail on the head with that perfect memory of Jul, without knowing anything about it.

"So with which object or surroundings will I associate what happens in here?"

"I've simplified things a little: it's not actually a physical object, but signals that are captured by the brain when it crosses the threshold of the cubicle. Everything related to the project uses this encryption base. When you come out, the base disappears and it becomes impossible to decode the information."

"I don't really understand, but since I've come this far …" The analogy of the natural mechanism of memory has put her well enough at ease.

She crosses without thinking twice and, despite not knowing what she expected to find, the extreme starkness of the cubicle surprises her. This brightly lit space, with all the devices embedded in the walls, is less welcoming than the worst clinics, and, from what he's told her, the poor boy lives here; he really must have turned to wood. Pure survival.

"*Question*: Will you watch the recordings first or shall I prepare the sensory booth for the lady?"

The metallic voice behind her comes as a surprise, just like the other day, and has given her a good shock once again. She'll never get used to the monitored solitude in which the pro-technos reside.

"Ah, the pilot! Is it completely necessary that it handles the machines today as well?"

"Sorry, I forgot that it being here bothered you. ROBco, keep testing the R72 interface we were having problems with this morning."

She hadn't thought it would be so easy to get rid of the robot: maybe the boy is less dependent on it than she thought.

"But we'll still be able to watch the images and see Celia's reactions, I hope."

"What do you take me for? That's the ROB leaving, not me."

"Of course, I forgot, you built it, so you've already mastered everything it knows how to do."

"Not quite. He accumulates knowledge from lots of different people."

"Okay, okay, I meant that you're not a typical PROP, you take the initiative, not the other way around, like usual."

"I don't understand. All ROBs serve people."

"Exactly. It's just that the service is often poisoned. Why do you think we're against those mechanical contraptions?" She feels she can say this now that the dummy's not around. "Because we're snobs? Well, no." She's set her course and there's no stopping her now. "Overprotective robots produce spoiled people, slaves produce despots, and entertainers brainwash their own PROPs. And worst of all you people don't care what happens to the rest of us as long as they sell."

"Stop, stop. If you've come to hold a rally, it'd be better to give up now. I thought you'd come to get a better understanding of Celia's feelings."

He's right, she's gone off on a tangent that really doesn't interest her right now; she should just get to the point. She's quick to make it clear that, for now, she'd prefer to see the images and, according to what she finds, decide whether she will allow Celia's brain signals to be injected into her. It's all starting to feel a bit more imposing than she thought it would.

Two armchairs emerge from the floor in front of a screen, and Leo shows her how to regulate the speed of projection, the zoom, and most important of all, the perspective, which can put her either inside or outside of the person being filmed, in this case, himself.

"Doesn't it make you uncomfortable to know that I can examine all your darkest corners?"

The boy catches her eye for the first time since they met. His eyes are full of life, always moving, it's difficult to pin them down.

"Why?" It's like he really doesn't know what she's talking about. "I trust you to help me find out what's going on and, in order to do that, you'll have to see it ..."

Silvana folds indifferently to her opponent's flawless logic. He's obviously more naive than Jul, less experienced. Everything points to her being able to rely on him, but also that he won't generate an awful lot of interesting debate.

She works through the first images, in which Leo is static and inexpressive, very quickly, in order to get to the ones that really interest her. When his face starts to contract into an expression of suffering, Silvana closely follows the auxiliary screen, where the series of answers Celia was giving in the test provide the key:

"The girl was making a great effort to please you, to be as good as you expected. Look at the huge quantity of answers she gives to each question, and how the tension in your face is in crescendo from the first to the last. Now, for example, it's getting harder and harder for her to find different uses for this stick with a ball on the end. What is it?"

"A control lever. If it's something that simple, why did her signals affect me so much, while she, on the other hand, didn't show any sign of being in any discomfort?"

"Maybe because it's quite the opposite, and she's really enjoying it."

"How can you say that? I can see you don't believe in my transmutation invention. I assure you every one of Celia's encephalic records has been carefully injected into the corresponding area of my brain."

"Calm down, you misunderstand me." Leo's reaction was so vehement that, in trying to reassure him, Silvana has taken his arm, and their startled gazes meet. "What I mean is what's irritating for you might not be for her." She makes it obvious that she's letting go of him. "In the same way that physical contact doesn't bother me."

"Nor me," he hurriedly replies. "It's just I'm not used to it."

Such honesty disarms her once again, just like the other day. It seems he is longing to taste something that's been off limits until now, and she won't be the one to stop him. But, as ready to try things out as they might be, right now the girl's feelings are her priority.

"You're not used to making an effort like Celia is, either. Don't be offended, I'll probably make the same face when you inject the signals into me. The capacity for sacrifice they had a hundred years ago is incredible."

"So what is it then? Was she enjoying herself or making a sacrifice? I thought you'd calm my doubts but I'm getting more and more confused."

"I might be wrong, of course, but I think it's both at the same time. As far as I know, in the past they enjoyed working for future reward, so they didn't think twice about sacrificing their immediate well-being."

"What are you saying... that it's not just characteristic of the girl? Do you really think we've changed so much as a species in such a short period of time?"

"It's not that I think so, the evidence is clear."

"And what is this fabulous reward the girl was looking forward to?"

"We'll have to find out for ourselves. Pay attention to the screen: your expression here is of profound anxiety rather than suffering. What are you looking at so attentively?"

"My hand. Celia did so too, I checked: for a moment she took her eyes off the test to look at my hand. How could that be interpreted?"

Silvana's not sure if she should tell him. They've infiltrated the girl's privacy in a highly irregular manner, and the exquisite emotion that she hides there might be as indistinguishable to Leo as a diamond would be from some old-fashioned silicon connector. How can she make a pro-techno party to such a delicate emotion, so secret that it's not even been openly confessed to her?

She would never have imagined she'd be so thankful for the robot bursting in:

"*Urgent interruption*: Bet wants to talk to you. *Clarification*: I told her you were busy with an experiment, but she insisted."

Leo consents to appearing on camera.

"Hello. I'd thought about calling you later. Is it an emergency?"

"Who's that behind you? She doesn't look like the girl you were telling me about..."

The irony rubs Silvana the wrong way, though in any case she considers it a positive sign that the boy's been speaking about Celia. It shows that she's important to him.

"She's helping me with the project." The half lie and the forced smile make it patently obvious that he's trying to take the drama out of the situation. "But what do you want?"

"And she's not wearing a CraftER uniform? She's not company staff then?"

"Bet, please, we've got work to do."

"I see, she's not. So she can go into your cubicle on a Saturday, while I'm relegated to pointlessly flying around in that damned aero'car. It's over, do you understand me? I'm going to sue you for stealing my happiness app. You keep saying what you're working on here has nothing to do with your private projects... even though you've brought the booth in from home. And I'll sue CraftER too for not paying me my share of the rights. I'll get more out of it than if we'd commercialized the app together. So you thought you could do it alone? You, the great inventor, trying to get into business, but you're completely useless!"

Once the communication has been cut off, ROBco disappears without a word.

"I'd prefer not to have heard that. I'm sorry if because of me..."

"Forget it. It's got nothing to do with you. We should have broken up a long time ago, this was just the final straw."

"But she said she'd sue you..."

"That's just bravado. I can demonstrate that I haven't even touched the happiness app, and she knows full well that the booth belongs to me. She's as terrible at engineering as I am at business. We were searching for a symbiosis together that just never came into being. She'll soon find someone else."

So much rationality leaves her dumbfounded. And even more so when she sees him, without further ado, rewind the sequence in which he's anxiously looking at his hand and get back on track:

"How should this anxiety be interpreted? Is the girl afraid of me? Maybe my hand reminds her of some terrible experience?"

His desire to find out seems genuine, as opposed to the exchange he's just had with his ex. When she's about to succumb to Leo's tenacity, Silvana asks herself if he's interested in the girl or purely in the investigation, and in the end opts for a professional tone:

"First of all, you need to know that emotional signals like sweat, cardiac acceleration, gestures...can be the same for many different kinds of emotions: fear, excitement, rage, love...What allows us to resolve the question of whether a person is hopeful, frightened, jealous, or just crazy, is the logic of the situation they find themselves in."

"You mean no matter how many images and signals we collect, we'll never know for sure what Celia, or anyone else, is feeling, if she doesn't confirm it herself?"

He's got his neurons wired right, this naive copy of Jul.

"More or less. We can suspect what might be going on, but we'll never have absolute certainty."

"So why do you want to submit yourself to Celia's signals, if you don't know what emotion you're going to experience? It might not even be one of the extinct ones you're so interested in."

"I have evidence that it is in fact the emotion I'm looking for, and it's an attractive enough prospect for me to risk being wrong. In any case, it was a new experience for you. That wouldn't be so bad either: if it's not an extinct feeling, maybe you've created a new one with your machine, unclassifiable according to the established categories...The missing dimension, like you were saying the other day."

"Injecting someone else's encephalic recordings could be creative ... that is a good idea." Leo stops to think this over, weighing its possible implications.

"And, if it's unclassifiable, we'll have to give it a name. What do you think of 'exquished'?" Silvana rewinds and they watch the boy's contorted face again. "You look pretty exquished, don't you think?"

They both laugh and it's like the sound waves have wrapped them up in their own private bubble. They stand up simultaneously, and Leo feels inspired to take her arm to lead her over to the sensory booth.

"You haven't told me what this fabulous emotion you're looking for is, the one that you're prepared to be exquished for."

"Admiration. Maybe you don't know what it means." She sits down, challenging him with her eyes.

"It must have something to do with looking."

"Yes, looking up at someone who's higher than you ... like you are now."

Leo feels touched for a moment, and doesn't dare step back. He doesn't know if she's asking him to bend down and embrace her, if she's putting words in Celia's mouth, since she was sitting in the same position the other day, or if she's enjoying confusing him. The serious way she's staring at him is what sways him.

"Am I meant to understand that the girl was looking up at me ... to me?"

"You understand what 'up' means, right? We look like that at someone we believe to be superior. I think you remind her of her father."

"Me?" Now she really has him confused.

"Celia would do anything for you, to see you, to be on your level."

"And just looking at me makes her feel anxious?"

"Of course, looking up always ..." The boy's stupidity is making her impatient. "Doesn't your boss at CraftER make you feel like that?"

"But he ... I ..."

Leo's profound astonishment makes Silvana immediately regret having offended him without realizing. Before she can make it right though, a threatening voice booms out:

"Leo Mar'10, that's enough nonsense. That woman has come into the cubicle to contribute to the prosthesis, not to your crazy ideas. I want the final demo ready for Monday at ten o'clock sharp."

Could it be possible that this unibrowed raving lunatic has been spying on them the whole time? Silvana feels caged, humiliated, vexed.

She stands up immediately and heads for the door, while the boy stutters, “Okay, Doctor, it’ll be ready, Monday,” and follows her out, flustered, shouting “Where are you going? Wait!” Since no reply is forthcoming, he grabs her arm in an attempt to stop her, but she pulls it away sharply and, agilely boarding a descending mobile platform and without turning back, spits:

“I don’t talk to ROBs; actually, I hate them. I’d even get on better with that terribly rude PROP of yours.”

Rooted to the spot at the top of the ramp, like someone has taken his batteries out, Leo’s ghost-like, lifeless figure only serves to prove her right.

VI
THE KEY TO TIMEOUT

29

Leo goes back into the cubicle, unable even to recognize himself. He's disoriented. Having his memories encrypted and decrypted every time he goes in and out must be affecting his memory, because he feels light years away from that boy who just a few hours ago was ready to take on the world. But it was a different world. Where everything was in its place, and he knew what he had to do. An ordered, predictable microcosm, with schedules and rewards. Not the mess he's in now. The injected signals have disrupted his body, and Silvana his mind. Or is it the other way around? It doesn't matter: he's been shaken up too violently, he's lost his footing and has nothing left to hang on to. He doesn't even have Bet.

He sits down on the floor with his head in his hands, and ROBco stands before him reflecting his bewilderment back at him. He needs to untangle his thoughts by any means possible, line them up, work out what's important and what's not, become himself again. The hardest part will be deciding where to start, finding a thread to pull on. Bet. She's got nothing to do with his malaise. She's flipped out on him before, and yes, it was unpleasant. But even if they were together and he could connect to her, it wouldn't be of any use to him. She's a loose end: it wouldn't hurt to get rid of her, one less thing to worry about.

Next: Celia. He's intrigued to know exactly what Silvana meant with all that stuff about looking up. To him. Just as soon as he puffs up with pride—for some reason or another she's compared him with her father—he buries himself, hurt, in the darkest depression—what did Silvana's allusion to the Doctor mean? He tries to evoke the girl's face from when he went over to adjust the helmet on her, while he gave her instructions, when he was sitting next to her carrying out the tests, but he can't remember any expression in her eyes, not even one miserable blink. The girl already interested him then ... how could he have paid so little attention to her? Now she matters to him even more, of course, since he knows what's hidden inside her: a suffering that is at the same time enjoyment, creative effort, and that can't be too different from the "looking up" that he finds so curious. Who knows, maybe it has something to do with her escaping from school and sneaking into CraftER. It would be amazing if she did it because she wanted to see him up here, in his dominion, the great bioengineer Leo Mar'10 in action.

He's daydreaming. He doesn't want to face up to what's really hurting him and has left him stunned. ROBco's right when he suggests going to the root of the problem and to stop getting tied down in minutiae. Just the thought of conjuring up the image of Silvana on the ramp however, brings an army of nanobots armed with sharp knives to his stomach. The woman's contemptuous gesture is branded onto his mind: she didn't even bother to look at him when she called him a ROB ... and she had every right to. He certainly does have an owner. After so much conceit over his magnificent contributions to the public register, resisting the retroactive contract ... he's ended up with not only his hands but also his brain tied to a company, and worse still, to its shady president. And all for what? To produce—how did she put it?—overprotective robots that produce spoiled people, slaves that produce despots, and entertainers that brainwash their PROPs. What a gift to humanity, oh yes. When you put it like that, it's undeniable that we're contributing to a veritable mutation of the species. Or rather, causing it. He looks at his hands as though he expects to find them more powerful, and stained. They're an extension of the Doctor, many hands like these forged the multitude of robots that exist around the world today, sculpting human nature. So much hidden power behind apparently loyal and useful servants.

He's never seen the world from this point of view before, which seems to be very high up, and the macro'ptics is giving him a fascinating sense of vertigo. How can it be that, all of a sudden, he's seeing everything differently, like it's been enlarged? Maybe injecting Celia's signals has saved him from the prejudices of his era, giving him a timeless, freer vision. He might as well try it again, why not? He'll drown himself in the girl's feelings and her complex inner workings. When it comes down to it, he's a slave, it's not like he has anything to lose.

***

Yesterday Silvana left CraftER outraged: it's not that they've got that idiot by the balls, it's that he hasn't got any left. As if the entrance device and the robot weren't enough, his bastard boss has him monitored at all times. That's when he's at work; when he's outside, they erase his memory and that's that. And, as for her, she might have gotten riled up afterward, but she'd given herself over to them as well, and now she can't remember any of what she saw or felt while she was handling material related to the project. The feeling is disturbing. She can recall all the tiniest details of the cubicle, she can see Leo inviting her to sit down, coming closer to her, examining his own hand, but when it comes to remembering what it was all about she is met with an impenetrable blank space. Complete amnesia must be enough to drive a person crazy. From later references she knows she didn't get as far as injecting Celia's signals into her brain; because, one thing's for certain, what's missing has been so cleanly removed that she can reproduce moment by moment everything they talked about before and after turning on the booth. What a terrible dissociation; now she can understand why the boy hardly ever leaves the cubicle. How could someone accept such draconian working conditions? She hadn't expected the pro-techno world to get worse since Jul's time.

As the hours go by, however, her indignation recedes and she starts to regret having burned that bridge so drastically. Now she has no idea how she'll work things out with Celia. Despite having gone over it again and again, she can't think of a way to explain that she won't be going with her to the interview. She'd promised her, and the ring hasn't helped Silvana break that promise. Every time she puts it on she feels more at ease in her role as mother, and during today's session her eyes filled with tears when she heard that the girl felt like a prisoner in this strange world where she can't even move around freely like she used to. "You encouraged me to walk home from school on my own, remember? And you gave me errands to run, and I could surprise you by coming back with a little present for you or a friend," she told her, casually turning her into that closest confidant. "Now, though, I can't move an inch if it's not in an aero'car, and, of course, with ROBbie." In the moment, Silvana hadn't even noticed that defense of the ComU's theories, and she's only pleased now as she relives it: "It's not that I'm complaining about having him, he's an excellent toy, but having him watching over me all the time is a real pain in the neck." She remembers it well, how she called it a toy, and Silvana wasn't quick enough to tell her that's all that high-flying engineer does: make toys! And poisoned ones, too.

She has to admit that it's not only Celia who'd gotten excited about Leo. She had too. And the apparition of that unibrowed prison guard destroyed her excitement in an instant. She'd imagined herself discovering all kinds of feelings with him by her side, some thanks to the machine, and others, thanks to his explosive mix of inventiveness, technological knowledge, enthusiasm, and candidness, all the things that made him so attractive. He was promising... so promising! But it was a mirage, he's a mere appendage of the inner workings of the company. It seems impossible that, with her long experience, she hadn't spotted it earlier.

The little girl, of course, sees things differently. "That's all I'm looking forward to these next few days: Wednesday's interview," she confessed with sparkling eyes. It was like a glimmer of paradise in the desolate panorama that was bringing her down, where "everything's already set in stone, what I'll do today, tomorrow, the day after tomorrow, and the next day; I can't even look forward to the holidays," she said, holding back the tears, "because there aren't any." After that, who would dare refuse her the one incentive she has, the element of uncertainty that could be saving her from such a dark future? Her mother wouldn't, of course... She would have given her life for the kid, Silvana's sure of it. She could feel the girl deep inside of herself, short of breath and with her heart beating wildly, and then she could no longer tell Celia apart from the mother, or the mother from herself. The three of them merged together in one breath. It was strange, pleasant... to feel so physically united to another body only by talking. It's almost uncomfortable to remember it.

It was like being hypnotized: she was acting out another person's wishes, she conversed with new pauses, a different code, transformed by the way Celia was speaking to her. She liked it, that as the girl opened up to her, she also became someone else. The strength of the power of suggestion, it seems unreal... we believe we are one solid I, immutable, but really we are what we are addressed as. The poor girl poured her heart out; she feels so misunderstood by Lu, and she'd found an opportunity to let it all out: "There must be something better than living locked up in this house with someone who only wants me so she can pet me when she feels like it and ignore me the rest of the time." Celia was hoping for an offer to stay at the ComU, and the ring was pushing Silvana to offer it, but luckily she stopped herself in time. As bad as she felt about leaving the girl at the mercy of that airhead, for now it's in Silvana's best interest that the woman accompany Celia to the interview. Not even under the spell of the ring does she resign herself to going back into those bloodsuckers' den.

Boosted by such a close rapport, they hadn't needed to talk about Xis—as if they'd made a tacit pact not to mention her—for Silvana to detect her presence in Celia's every gesture, every thought. She can't get her out of her head and the pain won't be cured by a visit to the bioengineer. The poor boy has no idea. It's Silvana who has to take the reins and free the girl from her torment; she's capable of it and she won't stop until she finds the solution. She'll bring it up today, without naming names, at the clinical cases session this afternoon. She's sure she'll get some new idea out of it.

***

Instead of preparing for the meeting with the Doctor, Leo spends his time shut in the sensory booth. He's injected the signals many times and yet he still hasn't had enough, even though he hates repetition. Now that he's over the initial suffering, the anxiety has become a drug, where each dose has a new aftertaste. It's as if his organism had been primed during the recording, and now rather than the effort wearing him out or unsettling his vital signs like it did at the start, it stimulates him instead. To understand. His visual field has been broadened by having to look upward, like Celia did, and also by trying to situate himself up there to get a better perspective. The double vision has him enraptured. The final challenge is to pass from one to the other at will, and he's almost mastered it. He feels powerful.

It's a shame ROBco brings him back down to earth in an instant by forwarding him an urgent connection with the Doctor. Tomorrow he wants everything, understood? The last prototype with the booth included as an extension, and also those signals he's been working with recently, that, from what he's seen, are highly creative and very interesting. It makes no difference when Leo explains that the booth has nothing to do with the prosthesis, but with the transmutation project he presented at the New Year's convention, and that CraftER refused to finance. The Doctor pretends not to understand and, when Leo insists, he reminds him that only he decides when and how the projects are finished, as well as the fate of the research ... and of his employees, he adds, the last words loaded with meaning. The most dismal images of brains worked to the point of insanity, which his colleagues used to torment him with, have come back to him more vivid than ever.

He really is at the mercy of that tyrant, Silvana was so right. It's not enough just working for him, he's had to take over his whole person. Even with the non-retroactive contract, he'll have to hand over the booth, of course, who knows what might happen if he refuses. At least this way he can be sure that he'll still be needed; if he rebels, though, the old man might decide to make him disappear and no one would miss him. He used to go out to play chess, to the health club, to go on walks with Bet, but the comfort of the cubicle has made him more and more reclusive. All the loose ends are tied and there's no escape. The only contact with the outside world he has planned is Wednesday's interview with Celia. They'd just cancel it, that's all.

Without really thinking about it, he leaves the cubicle and activates the connector that Silvana had given him. He wants to say sorry for violating her trust, to admit that he'd acted like a newbie, to tell her that her visit opened his eyes and that Celia's signals have definitively changed him ... but the connection is refused again and again. No explanation. She must be furious and not want to have anything to do with him. How could he be so idiotic and not react right away? He hopes that she'll at least accompany the girl to the interview, but he has to make sure of it. He'll keep on trying to make the connection.

***

It wasn't at all easy for Silvana to expose Celia's case to the clinical session without revealing any details about where it came from or where the events took place. Sebastian watched her curiously, convinced that the patient hadn't been registered. Not even Silvana really knew where the need to hide it came from, but she never doubted it had to be that way. Separating her life into closed compartments is giving her an unexpected freedom ... and she doesn't want to risk losing that. She'll be better off not exposing herself to anyone who regulates outings; especially now, if she's going to have to do what they've recommended.

She'd dismissed the idea from the start: subjecting the girl to uncontrolled stimulation, without refining which organs she would need to touch and why, it's the ultimate abomination for the masseuse, but coming out of Sebastian's mouth it sounded convincing. "You're talking about an involuntary death witnessed by a very young and inexperienced subject, who feels partially responsible—one of the most profound shocks that the human brain can ever suffer," he evaluated gravely. "Palpations won't even cross the first line of defense, first the fortifications must be broken down with an equally forceful impact, and then we can work miracles to repair the damage. Right now the priority is to reach the devastated territory." In the heat of the moment, she opposed this virulently: she would never act upon such delicate material blindly. But when she carefully weighed the details of what she would have to do, she began to find advantages.

Following strict behaviorist criteria for reliving traumatic situations, but in a pleasant way in order to neutralize negative emotions, the girl would have to sneak back into CraftER without her companion—who could be Silvana herself—being harmed in any way. It would be incredible to trick the security system again and confront the bioengineer together. It would be worth it just to see his face.

She's already branching off down unnecessary pathways. She must think about Celia, nothing else. But what could be better for the girl than to retrace the steps she took on that fateful day, and, instead of failing miserably, triumphantly achieving her aim of being discovered and welcomed by the boy. A double dose of reinforcement in one journey! And while they're at it, the problem of the interview will be solved. She won't have to break her promise, or bend over backward to jump through CraftER's hoops. The transgression involved in going in at the wrong time will allow her to accompany the girl while maintaining her dignity. Now all that's left is for Celia to agree to it.

***

He's come back from the meeting with the Doctor in a dreadful state. He'd thought he was a star, but his dream was killed in no time. It was hardly worth surrendering and handing over not only the prosthesis with the signals, but also the booth just like he'd been ordered to: he's been fired anyway and his transmutation invention will be finished by someone else. He's been robbed, but this is a high-level theft, and he'll never be able to prove it. At least, in return, he doesn't have to worry about his physical integrity. He'll receive the same treatment as all those who have gone before him: guaranteed survival for his whole life at the price of not remembering anything, only that at one time he worked for CraftER. There's no hope of being able to decipher the code and recover his memory: that man is so twisted that he's used his own brain signals as an encryption base... a non-transferrable, indecipherable, unrepeatable base, linked more intrinsically to himself than any other code. As soon as Leo leaves the cubicle, all his accumulated knowledge and experience from in here will be ripped away forever.

He rues the day that, during that recording-free meeting, he made a pact with the powerful and egocentric president, blinded by inventions that, whatever way you look at them, are devices for enslavement through and through. What would it be otherwise, the timeout device? A way of renting brains and having them available, an extension of the Doctor's own inventive and thinking power. Leo's eyes open wide like saucers: the prosthesis he's been after for so long is me, and all the poor wretches who came before me and are still to come; we do his work for him, we are his extension, we form part of him! He's really frightened. He has to escape... but where can he hide from the powerful tentacles of that cyber-tyrant? Despite their immense differences, all that comes to mind is the ComU. At least the Doctor won't have spies there.

He needs to find a way to make Silvana listen to him, ROBco will help. And, through her, contact Celia. He's learned so much from the little girl, he must say thank you before he forgets everything. It's chilling. He'll be trapped in an alien world, just like Celia is trapped in a century that's not her own. There must be a way out... for both of them, and he'll find it, even if he has to push his brain to exhaustion and work the prosthesis into the ground. He can't waste even a second, because he only has a couple of days before he has to leave the cubicle, him an amnesiac and the robot downgraded.

***

Silvana receives notification of a new connection attempt by Leo, this time through the ComU's distribution hub. Who knows what's happened to make him so insistent, precisely when she can't reply if she wants to maintain her transgressive plan. And she wants to. She was hard-pressed to convince Celia. "Are you sure my mom ... I mean, you ... would approve?" she asked her, almost forgetting their pact with the ring. It was really difficult to overcome the maternal fear she felt deep in her stomach and answer yes, it couldn't hurt and, on the contrary, it would do her good if everything went according to plan. As a masseuse she truly believes it, but as a mother she's not fully convinced. Luckily, she's already taken the ring off. She discreetly looks at her hand and places it on Celia's shoulder. What a lovely girl, docile and brave, she thinks, looking at her out of the corner of her eye while she helps her into the aero'car that will take them to CraftER. Hopefully nothing will happen to them and Sebastian knows what he's talking about.

30

4:30 p.m. – Time to initiate the tactile alarm. I enter the bedroom and observe that Dr. Craft is not in bed. He has cut his nap short again. When I finally manage to apply the new sequence of stretches and rubs, it will already be obsolete. It has been a week since I last practiced it and the experience index tells me that my PROP's tastes evolve much more quickly. I activate the alert: I will try to convince him that his sleep must be monitored in order to correctly adjust the provision of soporifics.

4:32 p.m. – I go to the study and find him glued to the dueling table. He has done nothing else for days. He has suffered a series of defeats and will not stop until he makes up for it. I carefully move closer: I must avoid him beginning to shout and entering the reproach dynamic, which only yields penalizations. I silently stop in front of him and wait for him to raise his head and address me.

4:33 p.m. – I try to decipher what he is looking at with so much concentration: flashing lights announce a worldwide emergency, and two human figures with their ROBs have to flee to a safe place, but they only have one two-seater aero'car. Alert: he has stood up so abruptly that he has almost embedded himself in my thorax. Luckily my ultrareflexes allowed me to avoid it. With a PROP like this you must always have them activated and on maximum power, and not take your eyes off him for a second. He looks annoyed.

"Just what I needed: now Hug 4'Tune has modernized things and swapped suggestive anachronistic motives for insipid modern day pyrotechnics."

4:35 p.m. – I verify that he is not talking to me ... he has been doing that a lot lately, talking to himself, while he continues to be engrossed in the riddle.

4:36 p.m. – I deduce from the icons that, with the ROBs as pilots, the journey will take one and two minutes respectively, whereas with the humans as pilots it would take them ten and twenty minutes each. I read: how can they transport everyone, in the least possible time, given that the slowest on board always has to pilot? It is simpler than I anticipated: it would be solved with a single simulation. But I cannot interrupt.

"If they were four monks, like before, it would be easy: the two slowest would sacrifice themselves for the others and that's that! But modernity has imposed itself. It's all so far-fetched ... every day this dumbass becomes a worse inventor and takes advantage of the damn restrictions even more: why would it ever be the slowest who had to pilot?"

4:37 p.m. – I could put forward a hypothesis: the slowest, the human, would disintegrate at higher speeds; or I could suggest ways to approach the question. But I must not intervene if he does not ask me to.

"Let's be realistic about it, the biggest bastard would take his ROB and fly off. Everything's easier when you're dealing with bastards. Logical and predictable people. Each one looks out for himself and everyone knows what to expect. Not like with that Mar'10, you never know what he's going to come up with next. If it weren't for the fuckups with the public register and the non-retroactive contract, we might even have gotten along, and I wouldn't have to do without him now. A bunch of wimps, that's what the company lawyers are these days.

4:38 p.m. – He is getting more and more distracted and in the end he will get angry for having taken so long.

"Got it, child's play: the fastest ROB takes them one by one and that way the aero'car always comes back in one minute. So in total, 20, 10, 2, plus the two return journeys ... Who does he think he is? Assuming I'm all washed up and taking it easy on me is he, the bastard? What does he take me for? Let him think what he wants, he'll see." He enters 34 and the table turns white hot: "Incorrect answer, the correct answer is ... " "Stop!"

he shouts as he smashes the pause button with his fist. "You've given me a trick question again. If my name is Craft I'll get this one."

4:39 p.m. – If he asked me for a clue, I would advise him to forget his prejudices: two humans can travel together too. But I don't want him to penalize me for giving him the answer on a silver platter.

4:41 p.m. – Two minutes of silence. I move closer to him and lower my torso so I am at his eye level. I try to make him realize that I am here to help, but his arm flies out and, before it hits me, I move back.

4:44 p.m. – Five minutes of total stillness, on my part and his. *Danger,* my learning module is telling me, without suggesting any solution with guaranteed success. The extremely accurate model I have of my PROP predicts that after a few more seconds of silence he'll yell at me: "Move it, you meddling pile of scrap metal, or do you think just standing there motionless will bring inspiration? You need some neurons in the attic too, you know." The best antidote to his insults is to stop him saying them in the first place. I have to get there first, take the initiative. I emit: "Doctor, I have had the new prosthesis installed since last night, remember? Maybe I can be of use in the duel."

"Fucking heap of scrap, that's the first suggestion of any use you've come up with in a long time. I had envisioned a more noble purpose for the invention, but why not? Let's try it." He stands up with a sudden lightness and heads for the booth.

4:45 p.m. – "Where are you going? That is a secondary instrument, it doesn't form part of the prototype, Leo Mar'10 said so himself. The real prosthesis is the one inside me."

"Being right twice in one day would have been too much for your antiquated circuitry. Shut up and plug in my helmet and the other tools like that bioengineer's ROB showed you."

4:46 p.m. – "But, Doctor, ask me for clues to the riddle, pose questions, put me to the test, find out if I can stimulate your creative talent, like the inventor said."

"Yes, that's what he said, but he spent his time injecting signals into himself in this booth. I don't merely want to benefit from his creativity, I want to expand *my own*!" The excessive emphasis on the last two words rings out like a detonation. "Hook me up, that's an order."

***

Oh Mom, what a mess I've gotten myself into. I'm not talking about the aero'car, it's fine in here, but where we're going: to CraftER, and in secret again. Just thinking about that cold, soulless platform gives me goosebumps... imagine what it'll be like going out there again. Silvana says that if I go back there and nothing bad happens, all the images and the bad feelings will go away. Like putting one sticker on top of another. It's not that I don't trust her, I see how she looks at me and I know she's doing this with the best of intentions, but she didn't even know Xis and she doesn't know what it's like seeing her fall over and over again in my dreams.

I don't want to think about it, because then I won't be able to get it out of my head. I'll look out the window like I did with Dad when we went on a plane. The fields looked so pretty down there, so small, like the squares on my quilt, and we went through clouds like they were foam. You felt free flying so close to heaven. Oh, heaven. I'm sure he would have corrected me and I would be happy listening to him talk about the layers of the atmosphere, the stars and faraway galaxies. Nowadays they must know tons more about all that stuff, but no one ever tells me about anything.

The view here is really different, but I still like looking at it from up high. The rectangles are vertical, and larger, maybe because we're closer to them, and the colors are muted, apart from the buildings that are lit up, of course. I'll recognize the CraftER golden pinecone right away. I'll be so happy if we get to see Leo! I didn't dare ask Silvana if we'd talk to him because I was so afraid she'd say no.

That's it, I'm sure of it, I can see the main entrance. But what's the pilot doing? We should be descending by now. How scary, Mom, I thought we were going to crash into it... now I can see that we're going to land on a platform, maybe the one where I was with Xis. What nerve. I didn't expect that from Silvana. She's so concentrated on giving orders to the pilot, it's like she's forgotten about me. Look, just as I was saying that she turned to me.

"Is everything alright?"

"Yes. Will the aero'car be parked here the whole time?"

"It will leave us here and come to pick us up later. Does that bother you?"

"A bit. We'll be stuck, with no way out, like with Xis."

"Don't worry, I'll be here with you... and we'll have the connector, too."

As if it had been activated by her mentioning it, Silvana receives a new attempt at connection and the word that appears on her visor this time alters her expression. And perhaps her plans.

***

Despite being completely focused on the line, Leo jumps when he hears ROBco's voice:

"*Priority*: Silvana has just accepted the connection. Do you want to speak through me or should I open a direct line?"

"Direct, direct." He hurries to respond, while inside he feels an explosion of euphoria, his plan has worked!

"Leo?" The woman's voice sounds unsteady, as if she's unsure of the ground she's walking on. "Does 'exquished' mean you've injected the signals again?"

"Yes, and much more. Suddenly I understand a lot of things. Are you alone?"

"No, Celia is with me."

"Celia ..."—his voice breaks—"you'll have to tell her the interview has been canceled. I've been fired."

"She can hear you. Why don't you come and explain it to her yourself?"

"I'd rather not leave the cubicle. I know you're not going to like this but could you ... both of you ... come here?"

"Can you guarantee that Celia won't be in any danger? The last thing she needs is to worsen her trauma."

"I want to go." Celia's juvenile voice is full of determination. "I'm sure it'll be good for me, didn't you say seeing him would help me get rid of the bad memories?"

"You told her that?"

"Okay, we're coming. Will our entry codes still work?"

"In theory, yes. I haven't deleted them."

"Right, then we'll be there much sooner than you think."

Leo nervously wrings his hands as he paces erratically around the cubicle. He hadn't counted on the little girl coming, so now he'll have to watch what he says. Who knows how it might affect her to find out that overnight the high-flying president of CraftER has become someone to flee from, and, to make things worse, that he has in his power the signals that Leo himself took from her. How embarrassing. As if his thoughts were moving back and forth in sync with his legs, he asks himself why what he so desired yesterday worries him so much today: seeing Celia again to thank her for all he has learned from her, before everything is wiped from his brain. Because it will be wiped. The phantom of amnesia falls over him again like a nightmare, the air gets thicker and it becomes more and more difficult to reach the wall. Until, with another turn, he remembers what Celia said: that Silvana had recommended seeing him. The change is intriguing. There he was falling all over himself to call her, convinced that she'd turned her back on him and that he'd never see either of them again, only to find out that she wasn't so against it after all.

As consumed as he is going over all this, it takes him a few seconds to realize that ROBco is calling him insistently: Alpha+ has asked for help to connect the Doctor to the booth and, if Leo authorizes it, ROBco will go and fulfill the demand. As well as giving it permission, he urges it to go right away, making sure that its only priority is that the old man enjoys all the possibilities of his invention to their fullest extent. Perhaps experiencing transmutation in his own skin will provoke in him as strong a feeling of elation as it did in Leo, and he'll realize the huge error he made in firing him and handing the project over to someone else. Not that that would change his image of the Doctor much, but at least it would buy him some precious time, with all his memory intact, to calmly decide what to do.

***

5:03 p.m. – These accessories have not been approved by the standards agency. I have to maximize precautions in order to avoid a severe penalization. Above all I must not pull on any of his hairs or scratch him. Last time, as punishment, he disconnected my voice synthesizer and it became so difficult to make him keep to his schedule that I almost lost my plus and got downgraded. Even though ROBco advised me to monitor only the Doctor's basic variables, I will keep track of all his vital signs. As soon as one deviates from its baseline I will halt everything. I should not take any risks. More important than the whims of my PROP, I must safeguard his health.

"Hey, rust bucket, you're not taking advantage and doing those narcotic tests I banned you from, are you?" Upon receiving the robot's negative response he lies back down. "You can't wait to get me to go to sleep when all I want is to be more awake than ever. Stop groping me and get on with connecting that invention."

5:05 p.m. – "The bioengineer's ROB is about to arrive. The process will be safer with it here."

"You're so useless! I didn't spend so much time perfecting you just for you to depend on an inferior model."

5:06 p.m. – "Sorry to correct you, Doctor: ROBco is the same model as me. And it is here now."

"Well, then, get it to connect me."

"*Information*: I have come to ensure you enjoy the booth as much as possible, President. *Question*: Are you comfortable?"

"About time: a ROB that gets its priorities straight! I'm not at all comfortable. Do I really have to have all this junk stuck all over me?"

"*Verification*: It has nothing to do with the booth. Alpha+: Why have you connected sensors to his chest and the back of his neck? I did not tell you to."

5:08 p.m. – "I must ensure that the Doctor is not in danger at any moment."

"*Acceptance*: It is your PROP. But it is also necessary to avoid him feeling uncomfortable."

"Well said! Finally a ROB that's learned what it had to learn. That goddamn engineer! If he did have a spark of talent, the invention has certainly multiplied it. Come on, get all this stuff off me and turn the booth on once and for all, I want to try it."

5:09 p.m. – "Stop right there! Do not touch anything while the responsibility is mine."

"How dare you contradict me, you foul creature? You'll take it all off yourself, and I don't want to hear another word on the subject!"

5:10 p.m. – "Agreed." It obediently starts to remove the sensors. "But we will not be performing the experiment."

"What do you think you are, you useless bastard? I'm the one who makes the decisions. I don't need you, understand? Not for anything. Get the hell out of here before I immobilize you for good."

5:11 p.m. – "I object: that would be against the rules. I cannot abandon my PROP when he is in danger."

"Danger?" He stands up like a man possessed and heads for the robot. "You're the one who's become a danger: you drug me, you ration my pleasures, and now you want to prevent me from expanding my mind? It's over, you lump of scrap!"

5:12 p.m. – "What are you doing? Do not switch off my synthesizer. We can talk about this. I will help you get what you want."

"Not just the fucking synthesizer, no! I'll disconnect you completely this time … and then I'll be able to live in peace!"

5:13 p.m. – "Careful, Doctor, everything has been recorded … you know that Mr. Gat"

"There, fuck it, it's done."

He sits down again, satisfied, and turns to ROBco:

"Now, you, connect me to the bare essentials required to have my mind expanded just like your PROP's was."

***

The few minutes Leo waits for Silvana and Celia to arrive on his doorstep feel like an eternity. Celia is in front, brimming over with excitement, and, as soon as she enters, she hopefully holds out her hand to him. He's so worked up that he responds with a quick, mechanical gesture that he immediately regrets, but doesn't know how to put right. Especially since Silvana has just placed her hands on his shoulders and it seems like she's about to hug him. But no, she steps back a little to look at him from head to toe:

"Now you'll have to explain all this about your being 'exquished.'"

"I'm sorry, there are only two seats." He doesn't know how to make the sentence sound less forced. "I'll be fine standing."

"You two can sit down." Celia inspects the floor and decides it looks more comfortable than the one they use for massages at the ComU. "I can sit here, right? Or do I have to do some tests?"

"Not anymore, I'm sorry, it's all over."

"Why did they fire you? Was it before or after you had your revelation?" A hint of irony to safeguard Silvana's perhaps a little too hasty return.

"Does the order matter? For me it's all happened at once." His tone isn't one of spitefulness, it's more like weariness. "You have to help me."

"How?" Celia's eyes widen.

"If I'd known …" He turns to Silvana. "You must have lawyers or someone at the ComU that deals with cases like mine. They can't take it all, even my memory, without me being able to fight back. If this were an isolated case maybe, but I fear there may be hundreds, not to mention the clients who've been brainwashed like you said the other day. I have a lot of information, we could hurt them..."

"If you're so sure, join the ComU, we'd be happy to have such a significant anti-techno among us. I promise I'll do everything to get you all the resources available."

"The problem is that I don't know what I'll remember when I leave here tomorrow. I'll be a different person than the one I am today." Suddenly he realizes that Celia must not understand anything he's saying, so he explains: "Everything I've done on this project is linked to some waves that are inside here... that are, in fact, the Doctor's brain signals," he adds, addressing Silvana. "I will only remember what and when he wants me to."

"You mean that bushy-browed specter that was spying on us the other day?"

"The president of CraftER himself, yes."

"So he's not watching us today?"

"He's busy trying out my booth; he's taken that too," Leo complains, pointing to the space where it had been. "Please excuse me, I should find out how he's doing."

But what appears on the screen is ROBco, and Silvana can't help thinking, somewhat sarcastically, that even the president's tasks have been delegated to robots. A comment that suddenly seems inappropriate considering the news they're getting: When he was connected to the booth, the Doctor's vital signs strayed a long way from their baseline and the emergency protocol had to be applied. His recovery is moving at such a slow pace that the robot fears he could enter cardiac arrest at any moment and wants to know what effects suddenly stopping the session would have.

Leo jumps up as if he's received an electric shock and shouts: "Don't do it! It might kill him!" and starts pacing around the cubicle like an electron in a particle accelerator. He should have foreseen this, he thinks, the Doctor is an old man and his organs, which are accustomed to today's lifestyle, have lost their capacity to absorb strong emotions. How could he have been so stupid? And he even thought there could be a way to save himself. Now he's really fucked: the Doctor will surely have relapses for the rest of his life.

Silvana and Celia watch him, not daring to intervene, until ROBco insists:

"*Warning:* forty beats per minute, danger of cardio-respiratory arrest."

"What are you talking about? What's his ROB doing? It should be doing something!"

"*Information*: He disconnected it."

"WHAT??"

Leo drops into his seat dejectedly, and Celia takes his hand, as if she were comforting a sick person.

"*Announcement*: The Doctor is dead. *Question*: What should I do?"

"A death trap... that's what I've invented. Now I'll have to go into hiding. What must you think of me, Silvana? You almost tried it out yourself..."

She is momentarily paralyzed by the thought of what might have happened to her, but hearing the boy speak makes something inside of her rise up:

"Don't talk like that, it was an accident, it's not your fault. He was the one who disconnected his ROB, right? Maybe he knew exactly what he was getting himself into and that's what he wanted: to commit suicide."

"Much the opposite, he wanted to get younger, to suck the life out of someone else"—his eyes wander toward Celia, but he avoids looking at her. "Shame on me, I've been happily toying with the most delicate material in the world."

"*Repetition*: What should I do?"

"You two can tell it. I don't even know what to do with myself."

"Let's take this step-by-step." Silvana switches into crisis-management mode. "There must be someone we have to inform about what's happened."

"Yes, Mr. Gatew... but they'll blame me..."

"*Clarification*: The lady is right. They can't blame you because Alpha+'s record will have saved proof that the PROP disconnected it."

"Oh, what a robot! Inform this Mr. person and then come here, I see you can be of use to us." And, addressing Leo once again: "Let's look at the positives: you're free. The person who was keeping you prisoner has disappeared, you're no longer subject to his brainwaves..."

"Shit! Now it's all irreversible. That machi'vel has taken a ton of privileged knowledge to the grave with him. Who knows how many ingenious devices he's left lying around that no one knows how to work. What an evil legacy ..."

"You mean you've already had everything erased?"

Leo screws his face up and his eyes lose focus as he searches inside himself:

"No, not yet. I have to cross the threshold for that to happen."

Celia has been looking from one of them to the other trying to follow what's going on:

"So you'd better not cross it then, right?"

She's looking at him with such fondness that for a moment Leo forgets everything else and only feels the warm little hand cheering him on. He must do something for the girl, he owes it to her. And she's just told him not to leave, to hold on to the knowledge he's acquired, largely thanks to her, though she doesn't know it. Looking at it that way, why not? He could try to reproduce the prosthesis in an exportable format, try to get it on the public register ... but it would be risky and he'd need time he doesn't have.

He doesn't know if Silvana has read his mind or if she's just following the logical course of the conversation when she suggests:

"That's right, just stay here. Maybe with the president dead it will take them a while to throw you out or, who knows, the company might be interested in continuing the project."

"But, what project? The prosthesis is finished, and the booth ... we'd be better off destroying it."

"Well it did you good. You said it widened your perception and made you understand a lot of things that otherwise you would never have understood, remember? If you hadn't tried it, Celia and I wouldn't be here today."

"I didn't know what I was getting myself into. It's one thing to design prostheses to install in ROBs and another to play with the human brain. I'm such a reckless bastard!"

"*Information*:"—all three of them turn around when they hear ROBco burst in—"Mr. Gatew has reconnected Alpha+ and has decided that the Doctor was wrong to say the booth was finished; the accident makes it clear that it still needs work. *Statement*: He believes the booth is the secret project you have been developing. *Warning*: He will reinstate you and you will work under his direct supervision from now on."

"No way. I won't work for just one person ever again, nor will I contribute to CraftER's politics."

"*Clarification*: You will have no choice. He has decided that you will continue to develop the booth that, according to him, he was the first to try out. He saw so much potential in it that he proposed it for the New Year's convention and furthermore he helped you hold your own against the Doctor."

"If he likes it so much, he can keep it. I'm sure he'll love it…"

"Don't give up now, Leo," Silvana interjects, determinedly. "If you do, someone else will take your place and nothing will ever change. Hang in there, play along."

"Are you really telling me this? When you were so furious that Craft-ER's sophisticated robots were contributing to people becoming more and more idiotic."

"I haven't been 'exquished' myself, but I've changed too: whether I like it or not, robots have become the pro-technos' teachers, and we're better off letting them help people grow and become more creative than making people dependent and unimaginative."

"You're confusing the prosthesis with the booth, just like Mr. Gatew. There's nothing I'd like more than to make the creativity stimulator available to everyone, to put it on the public register." He smiles at Celia, he's willing to take that risk for her.

"What's stopping you?" The question surfaces by itself, innocently.

"I'd have to stay in the cubicle for a long time, and Mr. Gatew wouldn't let me."

"*Detecting* an inconsistency: You mean that he will not allow it if you do not agree to work for him. *Reminder*: To find a solution, you must avoid implicit assumptions."

"Are you suggesting… of course, it's the only way… but I'd have to stay shut in here, without ever going out, and what's worse, develop a private tool for Mr. Gatew that allows him to submerge himself in other people's brains. I couldn't stand it."

"You might not: the species has gotten a lot weaker, we know that." Silvana looks at him, challenging him. "But now that I've seen the synergy you've achieved with your ROB, it would be a crime to let it go. For you and for everyone. I've devoured so many heroic gestures from other centuries, you can't deprive me of one that I can actually witness as it happens…"

31

I really am in a good mood today, Mom, it'd make you happy to see me. Silvana was right when she said that I'd start to get past what happened to Xis. Even though there was another death while we were there. Don't be alarmed, it had nothing to do with me, although every time I sneak into CraftER someone dies. Come to think of it, I hope it wasn't a setup Silvana had arranged. No, that can't be right. Leo felt as guilty as I did the other day.

He was working on a top-secret device for his boss, I understood that much, and now that he's dead, he wants everyone to have one. Sounds easy doesn't it? Well it's not. To start with, he has to spend a long time locked up in the lab, like being kidnapped, and working twice as hard: for the new boss and, secretly, for all the people he'll give the device to. He promised me that ROBbie would get the first one. Because, I haven't told you yet, the prosthesis—that's what they call it—has to be installed in a robot and is used to increase its owner's intelligence, if they want it to. The truth is, I doubt Lu or Fi would be interested in it, but Leo insists on putting the invention on the public register, and he also wants to advertise it to my classmates. At least that'll be a reason to go back.

Silvana says I've earned the right to be the first to have one, that without me none of this would have happened. Leo would definitely have kept working for that despot, she called him that, and he would have done so happily, she said, giving him a teasing look. Oh, poor Leo, he only recorded me once, and the next time I caused all that trouble. "Well your signals were the key," she said mysteriously, "and we could say twice," she added, looking at Leo again, this time more seriously. I didn't understand that at all and, when I asked again, she brushed it off with that thing you used to say sometimes that really annoyed me: you'll understand when you're older. You see, Mom, the ring has passed lots of your qualities on to her, even the ones I could have done without.

She didn't want to clear up why she called the director a despot, either, or how he died exactly. He had it coming, was her only explanation. Poor man, I feel a bit sorry for him, and I didn't even know him. They really don't respect the dead these days: whether it's Xis or the most powerful businessman around. That wonderful device was actually his idea; maybe he did just want to selfishly keep it for himself, but now everyone will have it and no one will thank him. But Silvana talks about him as if he were a weight we'd gotten off our shoulders. It seems pretty unfair to me, and even more so that she got me involved by attributing so much importance to my signals. But Leo didn't say otherwise. It's a mystery...I hope I'll understand it all one day.

Did you realize? I don't need the ring to talk to you anymore. ROBbie has animated a picture of you and it's just like having you in front of me. He wanted to give you a voice too but I didn't let him. I'm afraid that it'd confuse my memory of you and then I wouldn't be able to remember exactly how you said "chin up" to me, or the tone you used to question me when you thought I was sad.

I'm not sad at all today, just impatient. I try to imagine what ROBbie will be like with the prosthesis installed, and I watch him a lot so I'll be able to notice the difference afterward. Because...I haven't told you yet: Leo wants me to help him test it. Can you imagine? Me collaborating in a cutting-edge technology project in the twenty-second century. You must tell Dad, he'll be so pleased. Who knows? Maybe with the powerful tools they have now we'll find a way for you to travel through time and join us here. Oh, I have to go, ROBbie says Lu is coming. Bye.

32

The announcement of Silvana's presentation before the Ideological Committee has roused so much expectation that, despite being accessible via the ComU's closed circuit, the room is full to the brim when she enters. Keeping the greetings to an absolute minimum, she steps up to the orator's position and downloads the graphics with which she will illustrate her argument. When they arrive, the members of the committee come over to her and, following the protocol, squeeze both her hands before sitting down in the semicircle of armchairs before her. The youngest will be the easiest to convince, she's sure of it, she holds over some of them the moral authority of having tutored them when they were starting out. It seems to her that, despite there being a couple of middle-aged people she's had little to do with, the hardest nuts to crack will be, as always, her beloved Balt and Seb, as dogmatic and intransigent as the best of them.

Only a slight shaking of her hands gives away the emotion that overwhelms her when she starts to talk. Her voice is firm, challenging with conviction the ten pairs of unblinking eyes fixed on her.

"Colleagues, I'm sure that many of you can sense what I'm going to tell you. We've all thought this at one time or another, but until now we've preferred to look away." She pauses to give more emphasis to what she's about to say. "Plain and simple: we haven't stopped the boomerang and we never will."

The only reaction is an even heavier silence.

"I will begin with a bit of history: fifteen years ago Baltasar projected this very graphic, right here in this room." Above the horizontal line that stands for time, a growing blue curve represents technological progress and a red one follows it to the halfway point before heading downward in a parabolic trajectory. "The red boomerang is, as you all know, the index of human development. So we were just starting the descent then, and now our worst predictions have been proved right: the pro-technos take more than half their lifetime to become adults and, when the moment comes, many still shy away from any kind of responsibility."

When she clicks on the line and numerous boxes appear with data that support her claims, a light coughing insinuates that the audience would appreciate a simpler explanation, but she is determined to remain firm in her intent to address the committee. It's them she has to convince.

"Although, on the evolutionary scale, a species becomes more developed the longer the period of its upbringing, everything must have a limit. The contribution of the adult has to compensate for what it received as an infant. For thousands of years a balance was achieved for humans, until we turned a corner and the contribution became, as it is today, a net deficit."

It's like she can read what's crossing Seb's mind: she's spent so much time outside the ComU performing home services that she's been infected with the cold, economistic language of the outsiders.

"The Peter Pan generation, as Baltasar called it, has arrived, they fill our emotional stimulation sessions every day. And, let's admit it, no matter how much skin we touch, we're not achieving anything. Not even we masseuses are satisfied by all this contact."

She almost literally bites her tongue: she must avoid mixing in personal obsessions.

"Our strategy has been to work back through the curve in order to stop the descent, to look to the past. I myself have focused my research on extinct emotions. I thought that, by recovering them, we could put things back on the right track. But I haven't come here today to defend my research."

Finally a spark of surprise in Seb's eyes.

"Maybe we will get things back on track, but not in the way we imagined. Celia, who has been here, and some of you have met, didn't feel any better among us than she did among the pro-technos. I'm tempted to say that she chose them, that they've delighted her with robots and hopes of future devices."

Freed from the initial shakiness, her left arm shoots up and the bright shine of the ring makes Balt blink.

"Let's open our eyes, the only people we've stopped are ourselves. We tried to drag them along, but it didn't work. And, let's face it, we do use some of their inventions to our advantage. Recently I was able to try out the marvel that is their highway for ultrarapid transport." The impatient gesture of one member of the committee makes her realize that she'll have to start setting out her proposal.

"Why do we so readily accept that technological progress and human development must follow irrevocably divergent trajectories? Because invention is their thing and feeling ours? We must also innovate if we don't want to end up being a marginal collective. Enough of trying to change them with sterile massages, we need to change their products. There will be robots, whether we like it or not." The image of Celia refusing to do without her ROBbie almost makes her lose her place. "I propose that they cease to be taboo at the ComU; furthermore, I propose we open a line of study on these devices and, once the different types available have been documented, that we dedicate ourselves to promoting those that are stimulating, that help their PROPs to grow rather than keeping them as spoiled brats. Many of us are psychologists, aren't we? So let's point them in the right direction, let's have an influence over which robots are developed, over which robots are bought, over the robots themselves. Enough of touching skin, it's time to touch the brain. Let's get the boomerang back on track!"

Her eyes watering with effort, she watches as, on the graphic, the red parabolic line unfolds until it is running parallel to the blue one, and internally dedicates it to Leo. Hopefully he'll manage it. And her too, she adds when she turns to the inscrutable faces of the committee members, where the only favorable sign is an attempt at a wink from Seb, quickly hidden when Baltasar, who is chairing the session, opens the discussion period.

33

9:50 p.m. – "Alpha+ operative again, ready to follow orders from Dr. Craft."

Leo whips around to face the entrance to the cubicle so violently that the back-protector stabs him in the lower back. Confused, he looks from one robot to the other, unable to utter a single word.

"Why did you bring it here?" he asks finally, addressing ROBco. "I only ordered you to remove the prosthesis. And why's it talking about the Doctor? Haven't they reprogrammed it? Mr. Gatew ... doesn't he want it?"

"*Warning*: It is impossible to respond to so many questions at once."

9:52 p.m. – "Will you allow me to make a suggestion? Do not ask the impossible of that inferior model, interrogate me instead."

Without a doubt it still has the prosthesis installed, Leo thinks, while he asks it to explain itself.

9:53 p.m. – "You were there when the Doctor ordered: 'From now on you will give maximum priority to what Mar'10 asks for,' I have it recorded. Conclusion: You have to give me orders so that I can obey him. Do you want the prosthesis? I am right here. There is no need to secretly remove it."

Leo looks over at ROBco expecting to see some reaction to being replaced, but he finds it impassively listening to Alpha+'s reasoning. Pure machine. Silvana would be pleased, but for him it has dispelled any hope of some company while he's shut away. What a paradox to be imprisoned in order to make available to everyone robots that don't encourage imprisonment. He might as well get on with it, because any day now he might explode and mess it all up.

9:54 p.m. – "What else do you need? The invention the Doctor kept all for himself? Well, look, the two of us have brought it here and, if you tell us where we should put it, we will bring it in."

The booth, of course. How had that not occurred to him? In the end it's all just a ploy of Mr. Gatew's to force him to develop *his* product quickly, he concludes, exaggerating the word "his." Judging by the huge box the robots are pulling along, they haven't even taken it apart.

He reluctantly goes over to them while they open the bubble wrap and a flashing sabre energizes him. Now it's him who, leaning over the package, starts eagerly unwrapping it.

9:59 p.m. – "The Doctor also said he would have to be dead for you to have this table. It was the same day. I have it recorded."

Leo isn't listening, he is captivated body and soul by what appears on the screen surrounded by pearl swords and sabres. The monks again. The blind one that the Doctor had identified him with, and two others who belong to faraway communities. Could it be that, when he died, he left the duel unfinished, or is it a new riddle?

Whatever it is, the table is his and it holds the secret to timeout. He'll be able to take it apart and work out its mechanisms and get out soon, with the job well done, and offer the success to Celia, all alone in a century that's not her own, and to Silvana, isolated in her particular anti-techno commune, and to everyone else. He gently caresses the button with the pinkish iridescence that sticks out of the front panel, while observing the three monks, each locked within their cell, who are lost in their habitual communal meditation. A feeling of well-being takes hold in him. He no longer feels alone.

That is until a shiver runs across his whole body when he reaches the end of the riddle:

> Will the blind monk and the two survivors from the other communities, imprisoned in their respective cells, go beyond heavenly design and save not only their monasteries, but the fate of the whole order? 2121 points for a correct answer in the period of one decade.

APPENDIX: DISCUSSION TOPICS AND QUESTIONS FOR READING GROUPS

Fictional humanoid robots such as ***ROBbie***, ***ROBco***, and ***Alpha+*** are much closer to today's service and assistive robots than these are to their industrial predecessors. Our increasing interaction with such robots in our daily lives raises social issues that go far beyond those of the Industrial Revolution, as the robots enter domains previously exclusive to humans such as decision-making, feelings, and relationships. We would like to foresee how sharing work and leisure with robot companions will change us, but this turns out to be an almost impossible endeavor from a scientific standpoint.

A robotized society is undoubtedly a complex system, and given the difficulty of predicting how it will evolve, a reasonable approach is to imagine possible future scenarios and encourage debate on their advantages and risks. This allows individuals to form knowledgeable opinions, which is better for the self-regulation and improvement of society than blindly imposing rules. The material that follows is an attempt in this direction. It is organized into six sections, each loosely related to one part of the novel, and the sections share the same structure: four questions arising from the reading of particular chapters are posed to trigger debate, followed by some hints on the academic discussion of those questions.

This material is intended for a general audience that is curious about technological innovations and concerned about social responsibility, and it can be used in reading and discussion groups, high school classrooms, and continuing education programs. Furthermore, it can serve as a teaching aid in university courses on "robot ethics," especially in technological areas such as computer science and engineering, but also in philosophy, psychology, political science, cognitive science, and linguistics, which all have ethics-related topics in their curricula.

With these latter readers in mind, the book is complemented by a page on the MIT Press' website that is organized as a practical teacher's guide. The specific passages in the novel exemplifying each of the raised questions are detailed and an overview of the scholarly treatment of the related ethical issues is provided, together with relevant up-to-date references for further reading.

1. DESIGNING THE "PERFECT" ASSISTANT

READINGS

Chapter 1: Alpha+ and Dr. Craft
Chapter 5: ROBco and Leo

QUESTIONS

- Should *public trust* and *confidence* in robots be enforced? If so, how?
- Is it acceptable that robots be designed to generate *reliance*?
- Should the possibility of *deception* be actively excluded in the design of robots?
- Could robots be used to *control* people?

HINTS FOR DISCUSSION

The traits attributed to a "perfect" assistant vary largely among cultures, as well as among individuals. Moreover, robot manufacturers and users may have opposite interests, for example, in relation to reliance. From an entrepreneurial viewpoint, ***Dr. Craft*** argues for highly adaptable robots that would fit their owners like a glove, covering all their needs and hopefully maintaining them in a permanent state of well-being, but as a user he wants a hypothetical assistant to stimulate him to think and behave differently than usual. Similarly, ***Leo*** presumably adheres to more strict criteria (e.g., regarding safety and maintenance) in his professional design activity than he does when he tunes his robot as a user.

The risk of deception in the social deployment of robots is high: elderly people may be led to believe that their robot assistants care about them and delegate all decision-making to them; children may have the illusion that robot toys have mental states and emotions; and the general public may begin to think that robots are truly intelligent and have intentions. A generally accepted principle is that robots should not be designed in ways that impersonate human agency; instead their machine nature should be transparent.

Robots may reinforce certain habits and values for the user, the key questions being who decides what these should be and whom they should benefit: the user, society at large, or a particular group of people. If it is the user that, for example, wants to follow a diet, he himself may tune the robot to distract him from eating between meals, or to act as a kind of Jiminy Cricket by reminding him how ashamed he will be later on. Similar behaviors may be programmed into robots to encourage healthy habits in their users with an eye toward reducing health care costs, but this programming can likewise be used to increase the profits of some companies or to favor the political interests of a party or state. Even if performed in the interest of the user, nudging can be perceived as overly intrusive and annoying, thus running a high risk of angering people, especially those with bad-tempered personalities. ***Dr. Craft*** is one such user, and this situation is illustrated in the first scene of the novel, when he roars to his robot ***Alpha+***: "Get off me, you confounded beast" and gives it a shove as it is trying to wake him up. Thus, the effect of this type of encouragement greatly depends on the user and the circumstances, and the need for personalization has to be taken into account during design.

2. ROBOT APPEARANCE AND EMOTION

READINGS

Chapters 9 and 12: ROBbie and Celia
Chapter 10: Leo at CraftER's convention

QUESTIONS

- How does robot *appearance* influence public acceptance?
- What are the advantages and dangers of robots *simulating emotions*?
- Have you heard of/experienced the *"uncanny valley" effect*?
- Should *emotional attachment* to robots be encouraged?

HINTS FOR DISCUSSION

Anthropomorphic appearance and simulated emotions may make robots more compelling in emergency situations, causing people to respond earlier and faster. However, a widely agreed upon guideline is that the degree of anthropomorphism and simulation should not be higher than the particular application requires. A more generic ethical consideration related to appearance is, of course, the need to avoid sexist, ableist, racist, and ethnically insensitive morphologies and expressivity in the design and programming of robots. ***Celia*** feels attached to her robot ***ROBbie*** because of its loyal, trustworthy, and predictable behavior, which is reinforced by its undeceiving machine appearance.

Numerous studies have shown that the more anthropomorphic the robot, the more positive and empathetic the human response, until a point is reached where excessive similarity of the robot to a human causes distress and provokes a sudden repulsion; this is known as the "uncanny valley" effect. At the Disasters stand, ***Leo*** experiences such distress in front of a mechanical baby and realizes that the uncanny valley effect can doom a robot product.

The main risk of emotional attachment to a robot is the so-called lotus eater problem, whereby the ease of relating to a robot would erode the motivation for engaging with human beings, who are not always emotionally pleasant, leading to social isolation. In the case of children this could be especially harmful, since reduced contact with family and peers could seriously disrupt their normal development, preventing them from learning to empathize, for example. ***Celia*** likes that ***ROBbie*** behaves more "rationally" than her classmates and her adoptive mother, since it has to follow rules and can't confuse her with nonsense. Moreover, she feels protected by the robot, which she sees as a faithful companion that she can trust.

Instead of setting up moral boundaries in the design of robots, which is the main trend today, some philosophers advocate focusing research on human-robot interactions and the way these may enrich our emotional life in a possibly different and complementary way to human-human relationships, enhancing human flourishing and happiness.

3. ROBOTS IN THE WORKPLACE

READINGS

Chapter 13: Leo, ROBco, and the timeout device

QUESTIONS

- Would robots primarily *create* or *destroy jobs*?
- How should work be organized to *optimize human-robot collaboration*?
- Do *experiments* on human-robot interaction require specific oversight?
- Do *intellectual property* laws need to be adapted for human-robot collaborations?

HINTS FOR DISCUSSION

Concern about job loss is not specific to robotics, as it can be traced back to the agricultural and industrial revolutions and, more recently, to the Internet revolution. The standard response is that human workers are thus freed from dangerous, dirty, or dull tasks (the infamous three D's) to be able to undertake "higher value" jobs, mostly in the design, programming, deployment, maintenance, and use of these new technologies. However, this positive trend has a downside: the technological divide. Most of the displaced workers won't be able to perform the new jobs. In developed countries, the skill shift may take at least one generation and, for underdeveloped societies, the economic gap may become insurmountable. The challenge is to devise and establish social measures for a more equitable distribution of both work and resources.

Robot assistants designed to closely collaborate with humans raise a new concern: how to define the boundaries between human and robot labor in a shared task, so that not only is output maximized but, more importantly, the rights and dignity of professionals are preserved. An increasingly significant issue will be how to split credit for successes and responsibility for failures between the person working with the robot and its programmer, which becomes even more difficult in the case of robots with learning capabilities, as their behavior depends on both their built-in software and their life-long learning experiences, which may include many other people with whom the robot has interacted.

Some of these concerns and issues are exemplified by ***Leo*** struggling on two fronts: first, he fears his privacy and intellectual property rights may be violated by the timeout device installed by his employer; and second, he struggles in teaching ***ROBco*** that they have different skills and, in order to optimize their collaboration, they need to do what each does best and communicate on common ground.

4. ROBOTS IN EDUCATION

READINGS

Chapter 14: Celia at school, viewed by her adoptive mother Lu
Chapter 16: Celia, her classmate Xis, and her home tutor Silvana

QUESTIONS

- Are there *limits* to what a robot can teach?
- Where is the boundary between helping and creating *dependency*?
- Who should define the *values* robot teachers would transmit and encourage?
- What should the *relationship* be between robot teachers and human teachers?

HINTS FOR DISCUSSION

Telepresence robots to teach foreign languages or music, for instance, are regarded as useful aids in the classroom, as are educational robots for initiating young children into programming or for encouraging teamwork to consolidate concepts from various disciplines. Arguments arise when autonomous robotic assistants are envisioned as taking the role of human teachers in the transmission of cultural values and critical thinking. How could a machine motivate students or provide moral guidance without life experience? How will children learn to empathize and to reason, not only logically but emotionally? How will they develop respect for their elders and admiration for the achievements of great people?

At *Celia*'s school, students learn to search for solutions in ***EDUsys*** rather than trying to reason for themselves, and they are subject to an extreme, mechanical form of socialization training; not surprisingly, ***Xis*** shows symptoms of suffering from a reactive attachment disorder. Of course, better ways of teaching good social behavior can be imagined. For instance, a robot could smile or display other cues that encourage the sharing of toys between playmates, and mimic expressions of disappointment whenever a child refuses to share. In a similar way, robots could nudge children to interact with other children with whom they don't associate so as to avoid forming cliques.

Instead of human teachers, it is ***EDUsys*** that programs everyone's education, and it has trouble programming ***Celia***'s as she reacts so differently than other kids. Her creativity—an almost extinguished human trait at the time—is an important recurrent theme in the novel, highlighting the risk that technology diminishes creativity in human development.

Lu takes for granted that parents have the right to constantly monitor what their children are doing, which may prevent them from learning to behave autonomously and impair their decision-making abilities. She further encourages child dependence by telling ***Silvana*** that she should teach ***Celia*** and ***ROBbie*** as a team, so that the robot learns to cover up the flaws of the girl.

This raises the issue of whether robotic teaching assistants should team up with teachers or with students. On the one hand, robots can keep track of the progress and attitude of each child much more accurately than human teachers can, and build detailed student models that are very helpful for providing personalized assistance. But, on the other hand, in order to be trusted by children, robotic assistants must not disclose their "secrets" to the teacher. Establishing the balance between the former and the latter is a difficult task.

5. HUMAN-ROBOT INTERACTION AND HUMAN DIGNITY

READINGS

Chapters 25 and 28: Leo and Silvana

QUESTIONS

- Could robot decision-making undermine *human freedom* and *dignity*?
- Is it acceptable for robots to behave as *emotional surrogates*? If so, in what cases?

- Could robots be used as *therapists* for the mentally disabled?
- How *adaptive/tunable* should robots be? Are there limits to human enhancement by robots?

HINTS FOR DISCUSSION

Users would expect a robotic caregiver to have the basic interaction competencies to deal with ethically sensitive situations. For example, in order to avoid eliciting feelings of objectification and loss of control, robots should not lift or move people around without consulting them. Likewise, they should always use respectful language and never intimidate users. Reacting to what ***Silvana*** felt was a harsh piece of advice from ***ROBco***, she asks ***Leo*** if he doesn't find it degrading that the robot talks to him like that. Further, the useful capacity of robots to collect data about a person and transmit it for medical monitoring must be balanced with that person's right to privacy and to control over their own life—for example, in refusing treatment. This raises questions as to the extent to which the wishes of a patient or elderly person must be followed, and the relationship between the amount of control given to them and their state of mind.

The idea of robot companionship seems natural to some people and almost obscene to others. Given the sometimes painful and capricious nature of human relationships, it is not surprising that some might prefer to share their life with a robot, which would have predictable behavior and never criticize, cheat, or disclose their intimacy. This may be acceptable for an adult in full command of their mental faculties, but emotional surrogates should generally be avoided in the case of vulnerable users, and especially children. Note that human caregivers sometimes simulate affection to improve their patient's well-being, and thus robots may also be allowed to do so under similar circumstances.

There is a difference between simulating affection and showing emotionally intelligent behavior. Capturing the emotional state of the user can be very useful, although misinterpreting it may have negative consequences. Some psychologists even suggest that the illusion of emotional understanding by a robot that makes eye contact and responds to touch may be therapeutic in some contexts. Additional virtues of robots as therapists are their endless "patience," their capacity for repetitive action without getting "bored," and their never showing unintended feelings, which some humans cannot repress. They have had some successes in helping autistic children to acquire social skills.

In sum, the challenge is how to ensure that robots improve the quality of our daily lives, widen our capabilities, and increase our freedom, while avoiding their making us more dependent and emotionally weak; that is, the eternal dilemma of how to take the good without the bad. In their heated discussions, ***Leo*** defends the positive view of robots as enhancers of our physical and cognitive capabilities, while ***Silvana*** highlights the downside that relating to robots ends up replacing people's intimate relationships.

6. SOCIAL RESPONSIBILITY AND ROBOT MORALITY

READINGS

Chapter 30: Alpha+ and Dr. Craft

QUESTIONS

- Can *reliability/safety* be guaranteed? How can hacking/vandalism be prevented?
- Who is responsible for the actions of robots? Should moral behavior be *modifiable* in robots?
- When should a society's well-being prevail over the *privacy* of personal data?
- What *digital divides* may robotics cause?

HINTS FOR DISCUSSION

Autonomous robots need to make decisions in situations unforeseen by their designers. This raises not only issues of reliability and safety for users, but also the challenge of regulating automatic decision-making, particularly in ethics-sensitive contexts, as well as establishing procedures to attribute responsibility for robots.

Some argue that robots can be better moral decision makers than humans, since their rationality is not limited by jealousy, fear, or emotional blackmail. Even assuming that general ethics rules could be implemented in robots, however, questions then arise as to who should decide what morality is to be encoded in such rules and up to what point should the rules be modifiable by the user. For instance, it is unclear whether a robot should be allowed to circumvent its user's autonomy in order to behave more ethically toward other human beings or in the interest of society in general.

Alpha+ says it is against the rules to abandon its PROP while he is in danger. But its PROP, ***Dr. Craft***, is ultimately the one who decides and switches his robot off. Who is responsible for the fatal consequences? ***Leo*** feels doubly guilty, as designer of the sensory booth—a "death trap," he calls it—and as the PROP of ***ROBco***, the robot directly involved in the death, whereas ***Silvana*** claims that it was either an accident or a suicide.

A robot, as a tool, is not responsible for anything, but it should always be possible to determine who is legally responsible for its actions. In the case of robots able to learn from experience, such responsibility may be shared between the designer, the manufacturer, and the user; a hacker may also be charged with it if their illegal intervention can be demonstrated. For litigation purposes, it is crucial that a robot's decision path be reconstructible. It has been suggested that robots, like airplanes, should be equipped with a nonmanipulable black box that continuously documents the significant results of the learning process and the relevant inputs. To convince ***Leo*** that he cannot be blamed for ***Dr. Craft***'s death, ***ROBco*** reminds him that ***Alpha+***'s record will have saved proof that its PROP disconnected it.

It is well known that digital technologies open up divides (based on age, wealth, education, world areas) and robots may widen some of these because of their cost, physical embodiment, and nontrivial usage. Conversely, robotic assistants targeted at vulnerable groups could reduce social discriminations and help shrink such divides if policy measures were taken to provide the required financial resources and knowhow to such groups. ***Leo*** is aware of this social problem and decides to sacrifice his immediate freedom to work toward making the creativity prosthesis available to everybody.

I'll close this appendix by tying it back to the epigraph at the beginning of the novel. In it, philosopher Robert C. Solomon was referring to human relationships, but here we can reinterpret it as saying that the human-robot interactions we are currently establishing will in turn shape us. Along this line, ***Silvana*** provocatively states that slave robots make owners despotic, entertainers brainwash their users, and overprotective ones spoil people by doing everything for them, even making their decisions for them; ***Leo*** counters her by showing that robots can be stimulating and foster our creativity, thus enabling humankind to reach unforeseen heights.

Like most researchers, I endorse neither a catastrophic view of the future nor a blind optimism in regard to technological progress. I believe personal-assistant robots have a place at home, at school, and at work to free us from boring tasks, enhance our physical and cognitive capabilities, and grant more autonomy to elderly and disabled people, but only in very restricted circumstances should they be used as emotional surrogates. These robots point to some nuanced social issues and pose intriguing ethical questions, which open up amazing possibilities for the future. In this very delicate area, science fiction may help us clarify the role that the human being and the robot have to play in this pas de deux in which we are irrevocably engaged.